# Insects & People

CLYDE E. SORENSON, PH.D.

**Kendall Hunt**
publishing company

Cover image © Shutterstock, Inc.

**Kendall Hunt**
publishing company

www.kendallhunt.com
*Send all inquiries to:*
4050 Westmark Drive
Dubuque, IA 52004-1840

Copyright © 2017 by Clyde E. Sorenson, Ph.D.

ISBN 978-1-4652-7955-2

Kendall Hunt Publishing Company has the exclusive rights to reproduce this work, to prepare derivative works from this work, to publicly distribute this work, to publicly perform this work and to publicly display this work.

All rights reserved. No part of this publication may be reproduced, stored in a retrieval system, or transmitted, in any form or by any means, electronic, mechanical, photocopying, recording, or otherwise, without the prior written permission of the copyright owner.

Published in the United States of America

# DEDICATION

This book is dedicated to the memory of my mentor, Dr. Ron Kuhr, who saw in me the potential to be a good teacher.

# CONTENTS

|  | Introduction | vii |
|---|---|---|
| I. | Older than Dinosaurs: The History of Insects | 1 |
| II. | The Family Tree: Sorting Out a Million Critters | 9 |
| III. | Where Are the Bones? The Insect Exoskeleton | 19 |
| IV. | Eyes, Ears, Nose, and Throat: The External Anatomy of the Insect Head | 25 |
| V. | The Rest of the Story: The External Anatomy of the Thorax and Abdomen | 33 |
| VI. | Blood, Guts, and Other Innards | 41 |
| VII. | Have You Ever Metamorphosis? Insect Growth and Development | 49 |
| VIII. | Yes, Bees Do It: The Mating Game | 59 |
| IX. | It's a Bird! It's a Plane! Insect Muscles, Locomotion, and Migration | 69 |
| X. | Bugs, Bees, Beetles, and Butterflies: The Diversity of Insects | 79 |
| XI. | Better Communication through Chemistry: Insect Communication | 99 |
| XII. | Big Queens and Little Kings: The Social Lives of Insects | 109 |
| XIII. | Good Guys and Goodies: Insect Products and Services | 113 |
| XIV. | Buzzers, Biters, and Stingers: Arthropods That Cause Direct Injury | 123 |
| XV. | Guess Who's Coming to Dinner! Insects as Food | 135 |
| XVI. | Tick, Tack, Sick: Tick-Vectored Diseases and What to Do About Them | 141 |
| XVII. | Malaria, Plague, and Typhus: Diseases That Changed History | 147 |
| XVIII. | They'll Even Eat Fried Green Tomatoes! Insects and Agriculture | 155 |
| XIX. | There's a Fly in My Soup: Stored Product and Urban Pests | 165 |
| XX. | The Only Good Bug Is a Dead Bug! Insect Pest Management | 173 |
| XXI. | Pictures, Paintings, and Paraphernalia: Cultural Entomology | 185 |
| XXII. | The Hellstrom Chronicles: Insects in the Movies | 193 |
|  | Glossary | 197 |

# INTRODUCTION

How do you feel about insects? Do you hate them? Fear them? Have a mild interest in them? Never really ever think about them? Even if the last option describes you, insects affect you constantly, every day, in many really profound ways. Let's take a look at a typical weekend day, and see how often we bump into insects.

So, let's say you get up about 8 on a Saturday, and before you attack the day, you hop in the shower. The towel you use to dry off does a good job (and the cotton used to make it was partially insect pollinated). You then move to a nice, hearty breakfast—maybe some bacon, eggs, a bowl of cereal (sweetened at the factory with honey), with a handful of dried cranberries (pollinated by insects), some orange juice (pollinated by insects), and, of course a cup of coffee (pollinated by insects). Since it's also the first of the month, you go to the medicine chest, grab a tube of Advantage, and apply it to your dog Sam's neck, to protect him from ticks (and their diseases) and fleas.

You and your girlfriend have decided that it's a fine day for a hike, so you grab some trail mix (the almonds and cashews in it are insect pollinated) and make a couple of baloney and cheese sandwiches. The meat is in a plastic wrapper (partly to keep insects off of it), and it and the cheese were both in the refrigerator (partly to keep insects from exploiting them). You hop into the car and turn on the radio, only to be greeted by Brad Paisley's "I Want to Check You For Ticks." Not in the mood for country, you change the channel to the oldies station, and the Beatles are playing.

Arriving at the park, you prepare for your hike. You both apply repellent to your legs and arms to dissuade the mosquitoes and ticks from snacking on your blood, and your significant other applies lip balm (made with beeswax) to her lips to keep them from chapping.

On your hike, you pass through a meadow with abundant wildflowers, and you observe (and take smartphone photos of) the abundant butterflies flitting from flower to flower. You note, with some apprehension, that bees of various sorts are also abundant, but your girlfriend (whose dad is a beekeeper) assures you that foraging bees are harmless.

You arrive at a picnic area and claim one of the tables for your lunch. The picnic table you sit on and the shelter it sits in are constructed of wood pressure-treated with preservatives to keep termites and other organisms from eating it. Yellow jackets, attracted by the sugars in the soft drink you are drinking, start buzzing around, forcing you to cover the drink, remembering the time your brother was stung in the mouth by one of these wasps after it entered a drink can. A few houseflies also make an appearance and instigate a small pang of revulsion from you, since you have suspicions about where the flies might have recently been.

You get back in the car to go home. Van Morrison's "Tupelo Honey" is playing on the oldies station. Dusk is approaching, and you marvel at (and get irritated by) the large number of flies, moths, and even lightning bugs that are impinging on your windshield. When one particularly large individual smacks the windshield you tell your girlfriend that he probably doesn't have the guts to do that again, and she groans. Sitting at a stop sign, you notice the loud humming of cicadas coming from the trees around you, even though you aren't quite sure what is making the noise.

After a quick shower and a change, you and your girlfriend head out for a casual dinner. She is wearing a beautiful silk (caterpillar spit) blouse, and the lipstick she is wearing is colored with cochineal dye made from the crushed bodies of scale insects found on prickly pear cactus. She has a silver and turquoise butterfly-shaped pendant around her neck. The blueberries, melon, and strawberries in your fruit salad are all insect pollinated, as are the cucumbers in your girlfriend's house salad. You and she don't realize that, along with your salad, you are also eating insect eggs and aphids.

After dinner, you return home and turn on the TV to find *Men in Black* (with its giant roach villain) playing on a cable channel. While it's an awesome movie, you've already missed most of it, so you grab the remote and find *Antz* playing on another channel, and enjoy a cold beer (made with aphid-rich hops) while watching it. After Sylvester Stallone's brilliant performance is over, you decide to hit the sack. You brush your teeth, don your jammies, and go to bed. Sometime in the middle of the night, nature calls. You get up and stumble to the bathroom, and flip on the light—only to see a large cockroach disappear under the vanity. When you return to bed, you have a little trouble getting back to sleep….

Insects pervade our environment and our culture. In this book, we are going to try to understand who these animals are, how they function, and how they impact our lives in ways good and bad. Along the way, I hope you come to appreciate, at some level, how valuable they are, and how diminished our lives would be without them.

# Older than Dinosaurs: The History of Insects

The next time you visit a pond or a stream, study the dragonflies you see hovering over the water and zipping after each other in miniature dogfights. Notice how agile and adept they are, and then contemplate this—these creatures and their relatives have been dashing about like this for well over 250 million years. Insects, as a group, are very, very, old and this extreme age is one of many factors that contribute to their incredible success and their extravagant diversity. The history of insects and their relatives goes back beyond the time of land vertebrate animals, to a time when almost all life lived in the sea; it penetrates deep into the history of our planet. In this chapter, we'll briefly describe where insects came from, and we'll also look at the history of our attempts to understand and comprehend their astounding diversity and abundance.

In order to understand evolutionary history, we need to have a fixed reference point from which to launch our journey backward through time. Since evolution is generally a process that takes a very long time, evolutionary events are usually measured in thousands or millions of years, and scientists use a geologic scale based on big events in the history of life to simplify discussion of evolution (although it may seem to a beginning biologist that this time scale just complicates things). Figure 1.1 depicts a common construction of this time scale, and it's probably a good idea, if you continue your studies of biology, to get familiar with it. One of the most important things to recognize from this scale is that, for most of the history of our planet, complex animal forms didn't exist. During most of the Precambrian, which makes up most of Earth's history, life, if it existed at all, consisted of very simple single-celled or multi-celled organisms. Worm-like forms appeared towards the end of this period.

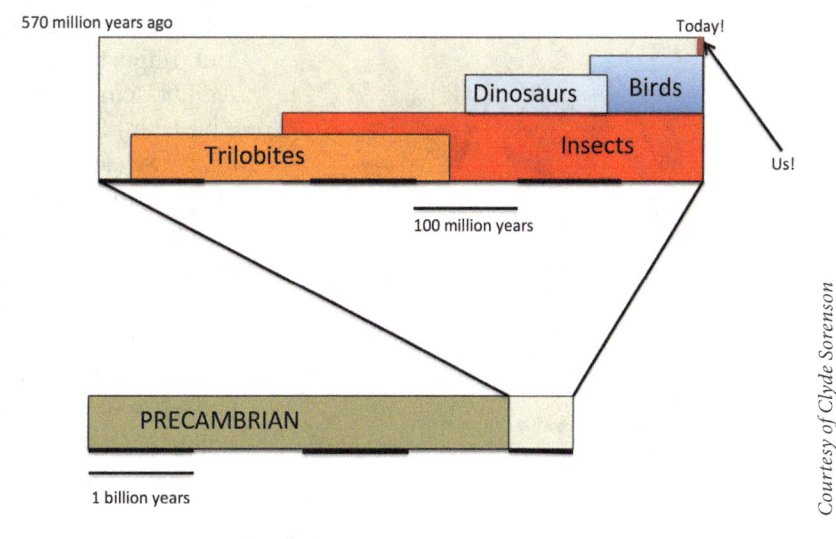

**Figure 1.1.** Geologic time scale of life on Earth.

The first of those major events happened in the Cambrian, with what is called the Cambrian explosion. Starting about 540 mya (mya=million years ago), animal diversity increased tremendously, and most of the major groups of living animals appeared, including the group that today contains insects. At this time, almost all life was in the sea, and most of it lived on the bottom. During this time, one group of arthropods came to dominance—the trilobites.

(From this time forward, major periods in Earth's history are defined by major extinction events—times when large numbers of species suddenly die out. The most famous of these mass extinctions is the one that occurred at the end of the Cretaceous, the one that was responsible for the extinction of the non-avian dinosaurs, but we'll talk more of that later. For now, it is important to understand that extinction defines, in a sense, the history of evolution.)

# Trilobites

The origins of the trilobites are not well understood, but these distinctive and important arthropods emerged early in the Cambrian, and persisted on the earth (well, actually, in the seas) for almost 300 million years. Trilobites (Fig. 1.2) are called "trilobites" because of the superficial organization of their bodies; when viewed from above, their bodies appear to be divided into three lobes. However, beyond that simple description, trilobites were incredibly diverse and filled a tremendous range of ecological niches. There were predators and scavengers, filter-feeding swimmers, and bottom-skimming detritivores. They ranged in size from less than an inch to over two feet, and many were armed with spines and protuberances that apparently protected them from predators (Fig. 1.3). Over 17,000 species of fossil trilobites have been identified.

**Figure 1.2.** Trilobite—extinct arthropods that once dominated the seas.

As I said earlier, trilobites were among the first arthropods to emerge, so this is probably a good time to discuss just exactly what an "arthropod" is. The phylum Arthropoda contains all the animals that have segmented bodies, bilateral symmetry (see Fig. 1.4 for what that means), jointed appendages (legs and such), and, importantly, chitinous exoskeletons. (Chitin is a really fascinating chemical that we'll discuss at length a bit later.) So, if we study one of the pictures of a trilobite, we can see (or at least infer) most of these traits. They clearly have segmented bodies, they are clearly bilaterally symmetrical, they appear to have jointed appendages, and they have an exoskeleton, and, upon microscopic examination, it looks like it is composed of chitin (although we can't be sure it IS chitin because of the fossilization process). So, they are "good" arthropods. We will discuss arthropods, their diversity, and importance later.

**Figure 1.3.** Trilobites with defensive structures.

2  Insects and People

Trilobites weren't static; there were "primitive" trilobites that occurred early in the Cambrian and more "advanced" and complex trilobites that appeared late during their reign. Their abundance and richness in the fossil record make them incredibly important to our understanding of other kinds of fossil animals since we can use the kinds of trilobites in a fossil deposit to date other fossils in the deposit. Trilobites dominated through much of the Cambrian but declined after that; they went essentially extinct altogether with the great extinction that marks the boundary between the Permian and Triassic periods, about 250 mya.

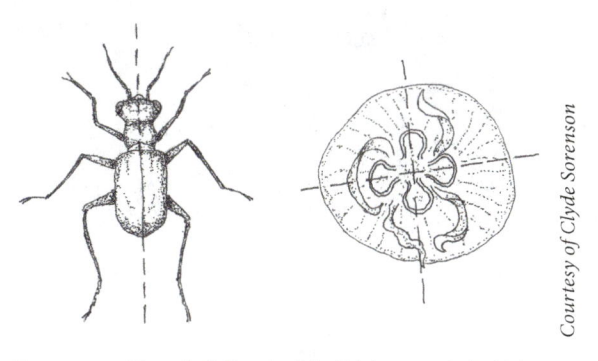

**Figure 1.4.** Bilaterally (left) and radially (right) symmetrical animals.

*Courtesy of Clyde Sorenson*

# Other Ancient Arthropods and the Rise of the Insects

Trilobites were among the earliest arthropods, but many other forms soon emerged. Also roaming the early seas were fearsome creatures called sea scorpions (Fig. 1.5); these were voracious predators that reached, at the extreme, almost 9 feet in length, and they were certainly the heaviest arthropods ever. They preyed on trilobites, early crustaceans (more about them later), and probably pretty much anything else they could catch. Early crustaceans appeared about 515 mya and rapidly diversified.

Horseshoe crabs (Fig. 1.6) appeared about 450 mya; these distant relatives of spiders have been virtually unchanged for essentially that entire time, although the modern species date to "only" about 20 mya. Because of the great similarity between the earliest horseshoe crabs and the modern species, many folks often refer to them as "living fossils," but what this really reflects is the durability of the horseshoe crab body plan and lifestyle. Horseshoe crabs are important to us today in medicine; we conduct extremely important assays for bacterial contamination with a component derived from their blue blood. Other species rely on them as well; some shorebirds, including, prominently, the red knot, exploit the eggs of horseshoe crabs (deposited on estuarine beaches) to sustain them on their northward migration (Fig. 1.7). Horseshoe crabs have been overharvested in recent decades for fisheries' bait, and the reduction in eggs on the beaches has consequently threatened the continued existence of the red knot.

Most of the other major groups of modern arthropods emerged by the Devonian, including the insects. The first insect probably appeared about 400 mya, although the oldest fossil insect has some characteristics of more modern insects, suggesting that the "real" first insect may have evolved even earlier. A huge problem in

**Figure 1.5.** Sea scorpion—extinct predaceous arthropod.

© AuntSpray/Shutterstock.com

**Figure 1.6.** Horseshoe crab, *Limulus polyphemus*.

© Kevin H Knuth/Shutterstock.com

Older than Dinosaurs: The History of Insects    3

**Figure 1.7.** Red Knots.

**Figure 1.8.** A springtail (Order Collembola).

pinning down exactly when the first insects came on to the scene is that they almost certainly were very small and soft-bodied, and therefore fossilized very poorly. (The fact that they were among the first terrestrial animals exacerbates this problem, since fossilization is rarer on land than in water, where burial in sediment is more likely.) The first insect-like animals resembled very closely some of the most primitive living hexapods: the springtails and proturans. Indeed, the oldest known hexapod ("six-legged arthropod") fossil appears to be a springtail (Fig. 1.8), one of the primitive hexapods we'll discuss a bit later.

Winged insects followed closely behind, perhaps as early as 380 mya, although again, the fossil record of the emergence of wings is very sparse and fragmentary. The importance of the development of flight in the insects cannot be overstated. Flight is a tremendous evolutionary advantage to animals, and insects were the first to do it. Flight allows insects to cover vast distances in brief periods of time; allowing them to migrate, to seek and exploit rare and scattered resources, to move long distances to find mates, and invade virtually every terrestrial corner of the globe. The ability to fly is one of the main reasons insects are, as we will see a bit later, the most evolutionarily "successful" organisms on the terrestrial planet, if success is measured by diversity and biomass. The exact origin of wings is a subject of much debate, and we'll get into this a bit later, as well. Some of these early flying insects grew to immense size by modern standards; some of the ancient relatives of dragonflies, the griffenflies, had wingspans of about 30 inches!

By about 230 mya, virtually all the major groups, or orders, of insects had emerged, and this is another key to the tremendous diversity of insects—their staying power. Insects have weathered at least four major and several minor extinction events—events that saw the demise of the trilobites, the giant amphibians, the archosaurs, the dinosaurs, and the Pleistocene mega-mammals. This impressive durability has allowed them to continually diversify, with periodic setbacks, for much longer than most other groups of animals. One of the main factors underlying the durability of insects is their size. A great diversity of small animals, inhabiting small ecological niches, is more likely to produce survivors during an extinction event than a much lower diversity of very large animals requiring extensive ecological niches. There are other reasons, but we'll address them later. As we've learned from amber fossils, some insect species have remained essentially unchanged for millions of years.

## The History of Insects on Another Time Scale

Insects have been around for hundreds of millions of years, but anything that even remotely looks like us has been strolling around on the earth's surface something less than a paltry five million years or so. We are late-comers on the evolutionary scene; if we are discussing our species, *Homo sapiens*, we've only been here a couple hundred thousand years at the most. But almost certainly, from the time of the first spark of human consciousness, we have been interested in insects in one way or another.

We can gain a great deal of insight into our early relationships with insects by observing our closest evolutionary relatives, the great apes, and in particular, the chimpanzees. Chimps spend a great deal of time grooming each other, and the primary targets of this grooming exercise (apart from the obvious pleasure it gives both participants) are the ectoparasitic arthropods (ticks and lice) inhabiting the groomee's fur. Famously, chimps are also insectivores; one of the most remarkable observations in the history of primatology (the study of primate behavior) was Dr. Jane Goodall's observations of chimps in the Gombe reserve purposefully constructing tools to "fish" termites and ants out of their woody nests. It is hard to imagine that our earliest ancestors didn't engage in these same behaviors.

While we don't have direct fossil evidence of our hominid ancestors' interactions with insects, our species has been leaving clues about its fascination with insects and their relatives since we walked on to the scene about 200 tya (tya=thousand years ago). Before we developed the written word, we were drawing, painting, and carving images of insect-like creatures on to and into rock surfaces around the world. Some of the 30,000-year-old art in the French Chauvet Cave depicts creatures that appear to be insects or insect relatives, and ancient rock art in Australia also depicts insect-like beings. The heavy reliance of recent hunter-gatherer societies on insects for much of their food (we'll talk much more about this in a later chapter) is another indication that we have had long-standing and intense relationships with the insects in our environment. We know that some of these same ancient cultures have a very profound understanding of the diversity of insects in their environment; they know which are good to eat, which are poisonous, which can sting, and which provide fibers. So it is probably safe to say that since we developed the abilities to communicate, we have been trying to make sense of that diversity.

Insects did not escape the notice of Aristotle (Fig. 1.9), the great philosopher and "first scientist," and he, in his *Historia Animalium*, attempted to describe where insects reside in the sweep of animals and to catalog their diversity. He had a great deal of insight into the lives and biology of insects; he understood that caterpillars gave rise to butterflies, and grubs to beetles. But, of course, his ability to discern distinctions was necessarily limited by the tools he had at hand and by the relatively limited ability to travel in his day (384-322 BC). Still, he did produce a catalog of insects in which he recognized about 47 different kinds.

One of Aristotle's students was Theophrastus (380-287 BC), who, among other things, was the Father of Botany, and was perhaps the first person to recognize, in writing, that insects can be serious pests of agriculture. He compiled a catalog of crop pests, thus becoming the first economic entomologist in history!

Pliny the Elder (uncle of Pliny the Younger) was a Roman philosopher and scientist who followed in the footsteps of Aristotle and attempted to expand on the knowledge laid down by him. (He's also famous for getting himself killed while making observations of the eruption of the volcano Vesuvius, as it swallowed

**Figure 1.9.** Aristotle.

Pompeii.) Like his predecessors, Pliny t. E. was a polymath whose interests spanned many disciplines, but one of his most noteworthy productions was his 10 volume, 37 book *Historia Naturalis*, his attempt to describe everything known about life on Earth as it was known in the First Century AD. Book 11 in this monumental work was devoted to insects, and much of this was devoted to the honey bee, an insect already long recognized as an important economic resource with fascinating biology. Pliny t. E.'s catalog of insects had grown to about 61 species in the 300 or so years between himself and Aristotle.

Sadly, not long after the time of Pliny t. E., scientific inquiry died in western civilization with the fall of the Roman Empire and was replaced by a societal preoccupation with religion. A scientific approach to natural history, based on observation of the natural world, was not fully revived until the Renaissance, some one thousand years later. During these Middle Ages, social upheaval due to invaders from the east, periods of depopulation (caused, at least in part, by one particular insect), and the Catholic Church dominated society. Contact with the Islamic cultures in the 11th and 12th centuries planted the seeds of the Renaissance, and by 1500 or so, natural historians and early entomologists began once again publishing their observations of insects in their natural world.

Our craving to identify and categorize persisted, and many of these Renaissance scientists attempted to produce catalogs or tabulations of the diversity of insects. In 1661, Johann Sperling, a German doctor and zoologist, in his *Zoologica Physica*, stated, "After all, we know at least 40 species of beetles, 50 of caterpillars, 70 of flies and over 100 butterflies," coming up with a total of about 260 species. There is a substantial problem with his tally, however; he counted caterpillars and butterflies as different species, while we know that many of the caterpillars were the same species as the butterflies they become. This presents a problem—just what is a species? We'll address that shortly.

John Ray, an English preacher, professor, and natural historian, together with his protégé Francis Willughby, developed a keen interest in the biology and diversity of insects during the late 1600s. They also counted larvae and adults as two different species, even though they were aware that sometimes a particular caterpillar turns into a particular butterfly. (Their reasons for doing this made sense at the time, because they were interested in describing the diversity of forms.) Eventually, Ray produced a work, *Historia Insectum*, which was published after his death, in which, based on his and Willughby's work, he determined that the number of insect species probably exceeded the number of plant species, and decided that, "…The number of insects in the whole world, on land and water must be in the region of 10,000…"

So, how many species of insects are there by today's measure? We'll tackle that in the next chapter.

## What Is a Fossil?

Fossils are, of course, basically relics of ancient living things. The word *fossil* means "of the earth," and suggests the origin of most fossils—they are dug from the ground. Fossils come in a tremendous diversity of types, but when we are discussing fossils of insects and their early relatives, they fall into two major categories: lithic, or stone, fossils, and fossils in amber. Lithic insect fossils form when an insect is buried in sediments, moved by either water, wind, or volcanic activity, very shortly after it expires. Such fossils need to be buried rapidly enough and deeply enough to limit microbial activity that decomposes living matter, although the chitinous exoskeletons of insects are pretty resistant to microbial degradation. Over time, additional deposits of sediments accumulate on top of the original deposit, and over millions of years, the sediment deposits are converted to sedimentary rocks by the pressure of the overlying deposits. Most lithic fossils of insects look almost like two-dimensional pictures of the creatures because of the tremendous pressures involved in converting loose sediment to stone.

Amber is fossilized plant resin, and not sap, as many people think. Sap is basically water containing the sugars the plant manufactures, along with other nutrients conducted by the plant's vascular system; the plant manufactures resins composed of combinations of complex carbohydrates to defend its self against insects and other invaders. Some of these carbohydrates are quite volatile (that's why pine smells like pine), while others are not. Some trees, both broadleafs and conifers, produce copious amounts of this stuff when they are wounded by an insect or some other kind of mechanical injury. If globs of these resins are covered by soil, and then deeply buried over time much like we discussed earlier, the volatile compounds eventually disappear, while the non-volatile compounds are chemically changed by the pressure, and often heat, resulting from being deeply buried. The end result is a hard, often transparent substance that has been prized by humans as a semi-precious gemstone. (The only gem that burns, by the way!) If an insect or other small animal was entrapped and engulfed by that resin all those millions of years ago, the volatile compounds in the resin might halt bacterial degradation and preserve the creature essentially intact, making that particular piece of resin prized by the scientists who study those creatures. Amber fossils often preserve the tiniest features of the insects and other life forms they contain.

In spite of the abundance of insects throughout most of the history of life on land, insect fossils tend to be somewhat rare due to their relatively small size, relatively soft bodies, and largely terrestrial lives.

# The Family Tree: Sorting Out a Million Critters

We left John Ray with his estimate of more than 10,000 species of insects in the world, but we know his definition of a species was somewhat different from what most of us regard as a species. So before we proceed on to our discussion of just how many insect species there are, we ought to settle on just what a species is.

This is not as easy a question to answer as one might expect it to be, and there is, to this day, still a great deal of debate over the definition among biologists, in spite of 2,500 years of scientific thought over the matter. Ray and his predecessors, and, for that matter, many who followed him, used a morphological concept of a species—all those individual organisms that look alike, are a single species. This explains, in part, why he and his colleagues often recognized the caterpillar as one species and the adult butterfly as another, even though they knew one turned into the other. Prior to an understanding of evolution and the spectacular and beautiful relatedness of all life on earth, this made sense. For most of these scientists, the world was created, as they saw it, by a divine hand, and they sought to understand diversity as a means of discerning this creator. Charles Darwin would dramatically change this worldview with the publication of his landmark book, *On the Origin of Species*, but even he did not have a grasp on exactly what mechanism would allow the natural selection he described to function. (More on this later.)

For most of the last 60 or so years, most biologists have used the Biological Species Concept, defined by Ernst Mayr more or less as "A population of organisms capable of reproduction leading to fertile offspring, and reproductively isolated from other such populations." This is a functional definition that works well with the theory of natural selection, even though, when Mayr created it, we still did not understand exactly how selection worked. One of the most important parts of this definition is that "reproductively isolated" bit- this usually means that there is some physical, behavioral, genetic, or morphological ("shape") difference, under the control of genes, that prevents members of different species from producing viable young or even mating at all. All human beings on this planet are the same species, because, even though it is highly unlikely that an individual from North America will have the opportunity to mate with someone from Asia, if they did, their offspring would be viable and able to reproduce with another human once they grew old enough. This is true for any two of us of opposite sex and appropriate age from anywhere on the planet. With respect to insects, however, it can be extremely difficult for us to discern between closely-related species, and to identify those reproductive barriers, and this is one of the greatest challenges to our ability to understand insect diversity. Things get really confusing when two closely related species occasionally do mate with each other and do produce offspring, and these offspring do have some degree of fertility.

Up until fairly recently, most of our efforts to identify and distinguish species were still based on morphology: the shape, color, and appearance of the parts of insects. The entomologists who make their careers attempting to describe the diversity of insects (we call them taxonomists and systematists) would collect specimens, and then spend great amounts of time carefully examining and describing them, to identify the sometimes tiny ways closely related species might differ in appearance that would allow us to distinguish them. This can be an extremely daunting task. Some very closely related species might differ only by the shape of a wing vein, or the arrangement and number of setae ("hairs") between the insects' eyes. In some cases, there is no consistent morphological difference between species and yet different populations might functionally be different species, based on behavioral observations.

Over the last couple of decades, however, systematists have acquired a powerful new suite of tools from the science of genetics that allows us to understand much more about the relationships between organisms. We can now study the similarities and differences among the actual genes responsible for the morphology we previously examined. Evolution, at work for hundreds of millions of years, creates branches and tangles that can be extraordinarily difficult to understand if we only stare at the end result; genetic tools allow us, to a certain extent, to actually trace the map of those branches. In some cases, evolution drives distantly related organisms towards similar appearance, due to similar ecological function. We call this "convergent evolution." In other cases, very closely related species may be driven towards radically different morphology as populations of one species diverge to exploit different ecological niches; we call this "divergent evolution." Genetic tools allow us to see through some of these tangles. However, there are many cases where genetic analysis raises as many questions as it answers; our best understanding of evolution, and the classifications we construct from that understanding, are probably going to be products of both genetic and traditional morphological approaches.

So, after that long and meandering discussion, let's get back to that question of how many species of insects there are by today's definition. Figure 2.1 is a pie diagram of the diversity of animal life on the planet. If you study this for a moment, you'll notice that insects constitute almost three-fourths of all animal diversity—there are just under 1,000,000 named species of insects! This compares to about 5,400 species of mammals, just under 10,000 species of birds, and about 32,000 species of fish. There are about 350,000 species of beetles (order Coleoptera) alone! Insects, by a very wide margin, have the greatest diversity of any group of eukaryotic (that is, living things whose cells have nuclei) organisms. (Insects far outnumber the known species of prokaryotic bacteria and archea, but we are just beginning to understand their diversity,

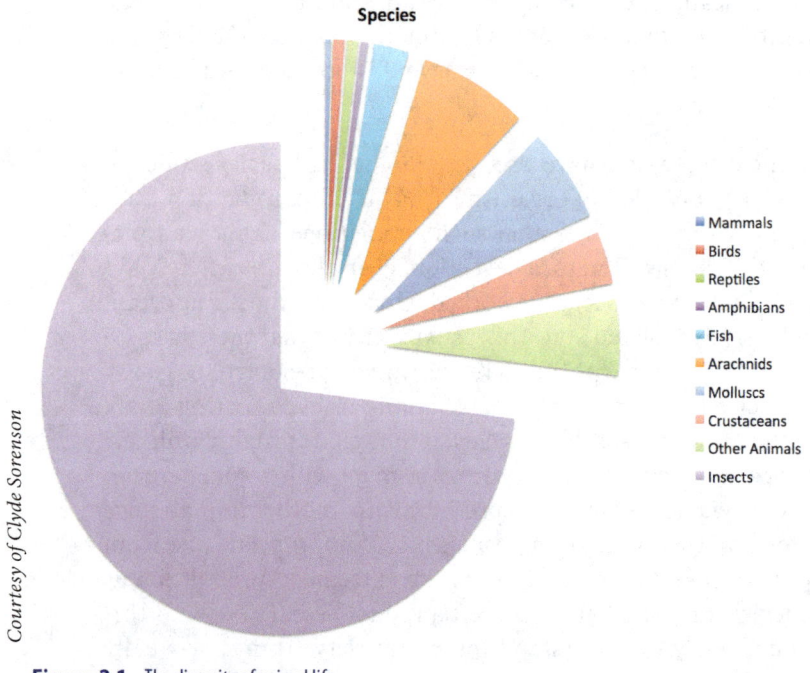

**Figure 2.1.** The diversity of animal life.

*Courtesy of Clyde Sorenson*

10   Insects and People

and we really don't know how to define a "species" for these organisms that reproduce asexually.) We will be discussing the diversity of different groups of insects a bit later.

Insects are not only incredibly diverse; they are also incredibly abundant. Insects, taken together, almost certainly outweigh us globally; in most terrestrial and freshwater ecosystems, they probably outweigh all other animals, sometimes by many fold. In eastern North America, an acre of hardwood forest could, in the right year, harbor half a ton of periodical cicadas alone!

# A Hierarchal Approach to Classification

Making sense of the diversity of life, given that there are over 1.2 million described species of living things, and perhaps another ten to thirty million yet to discover and describe, is an exceptionally difficult task. One of the most important goals in our approach to classifying living things is that our classification should demonstrate how each species is related to others, and to do this, we use a hierarchal approach to classification. In this system, we first describe big groups of living things with very fundamental similarities and then describe progressively smaller groups with more narrowly defined similarities (Fig. 2.2), until we get down to the finest resolution, the individual species, which we've (sort of) defined previously. In modern science, each species receives a two-part, or *binomial*, scientific name, assigned by the scientist who officially describes that species; the first part of the name indicates the genus the species belongs to, and the second part, the species within that genus. We can credit one remarkable scientist with these two incredibly useful innovations—Karl Linnaeus, a Swedish botanist who lived in the 1700s.

While Linnaeus's ideas about species and their origin were radically different from those of modern biologists, he did recognize that there were relationships between species and between groups of species, and while his motivations were different from ours, he did establish the approach to classification that we use to this day. He was the first to systematically use a binomial approach to naming species. This replaced a sometimes incredibly cumbersome system where each species "scientific name" was actually a paragraph-long description of the organism in Latin (Fig. 2.3). Linnaeus's innovation made clear, concise conversation about living things much easier. By the way, the bionomial name for a species is always Latinized (for linguistic stability) and therefore, italicized; the genus is always capitalized, and the species name (the "specific epithet") is not. Every

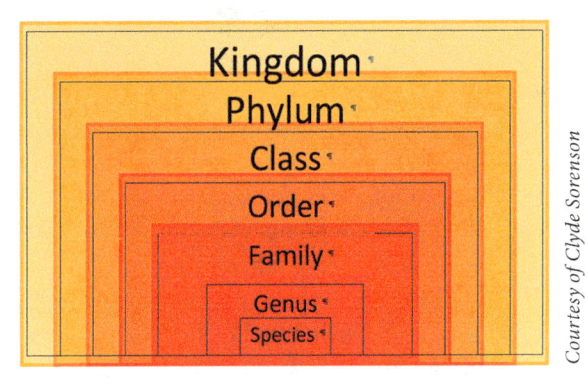

**Figure 2.2.** A hierarchal approach to classifying living things.

**Figure 2.3.** The scientific name of the periodic cicada before Linnaeus: *Papilio media alis pronis prefertim interorbis maculis oblongis argenteis perbelle depictis* (which, very loosely translated, means "The butterfly with a mark between its eyes and very pretty silvery wings") After Linnaeus: *Magicicada septendecim* (which means "The magical cicada that appears every 17 years")

The Family Tree: Sorting Out a Million Critters   11

**Figure 2.4.** *Helicoverpa zea*: the bollworm, tomato fruitworm, pepper podworm, corn earworm, etc.

scientific name is unique; a Chinese entomologist and an American entomologist who want to discuss *Helicoverpa zea* (Fig. 2.4) with each other can do so with no chance of confusion, since there is only that one scientific name attached to that species of moth. If these two scientists were to rely on the common names used for these insects, however, there is great opportunity for confusion. *H. zea* has about a dozen common names in the United States because as a caterpillar it feeds on many important crops. Depending on where we find it, it might be called a bollworm (in cotton), a corn earworm (in corn), a pepper podworm (in pepper), and so on. In China, the insect they know as the "bollworm" (well, whatever "bollworm" is in Chinese) is a completely different species, *H. armigera*.

Let's look at Figure 2.2 again, and talk a bit about this hierarchy that Linnaeus established. The largest, most comprehensive grouping is, of course, "Life" meaning all living things. Living things have complex organization, the ability to relay signals of one sort or another within their bodies, and can reproduce themselves through the genetic information they contain. Life is divided into several **Kingdoms**: the plants, the animals, the fungi, the protists (all of which are eukaryotic), bacteria, and archea (the latter two are prokaryotic—single-celled organisms without nuclei or other organelles). Insects belong to the **Kingdom Animalia**—just as we do. Animals are multicellular, eukaryotic organisms that are heterotrophic, that is, they acquire the energy and nutrients they need from other organisms (or things produced by other organisms), by feeding on them.

Within the Kingdom Animalia, insects belong to the **phylum Arthropoda**. We are going to investigate just what an Arthropod is a bit later. Here is where we diverge from the insects; we belong to phylum Chordata—all those animals with a backbone or something like a backbone. So, while we, and insects, are all animals, the differences between us and them are substantial, profound, and ancient.

Within the phylum Arthropoda, all insects belong to the **class Insecta**. The Insecta are defined as those arthropods with among other defining traits, six legs, three major body regions, ectognathous mouthparts (exposed mouthparts not hooded by the head capsule), and, in many cases, wings. We're going to come back to some other classes of arthropods a bit later.

Within the class Insecta are **Orders**. There are about 30 orders of insects. I say "about" because there is some debate among insect systematists about just which orders are valid. The species included within a given order generally share the same pattern of metamorphosis, the same kind of wings (if they have wings), and the same kind of mouthparts. Let's pick an insect to use as an example, say, the Japanese beetle (Fig. 2.5), from this point on. The Japanese beetle belongs to the order Coleoptera—all those insects with complete metamorphosis; forewings modified into hard, protective, unveined structures called elytra; and chewing mouthparts.

**Figure 2.5.** The Japanese beetle, *Popilla japonica*.

12   Insects and People

| KINGDOM | Continent | Animalia | Animalia |
| --- | --- | --- | --- |
| Phylum | United States | Arthropoda | Chordata |
| Class | North Carolina | Insecta | Mammalia |
| Order | Wake County | Diptera | Primates |
| Family | Raleigh | Muscidae | Hominidae |
| Genus | Hillsborough Street | *Musca* | *Homo* |
| Species | Mitch's Tavern | *domestica* | *sapiens* |
| Common name | The bar from Bull Durham | The house fly | Us! |

*Courtesy of Clyde Sorenson*

**Figure 2.6.** Your address, in a hierarchal system of classification.

Within the Order are **Families**. Our Japanese beetle belongs to family Scarabeidae. (Family names almost always end in "-dae.") Scarabeidae is defined as all those beetles with stout bodies, clubbed, fan-shaped antennae, and "C" shaped larvae. There are over 40,000 species of scarabs in the world. Many are important pests of agriculture, and some have had great religious importance in some parts of the world.

Within the family are **genera** (singular **genus**). Our Japanese beetle belongs to the genus *Popillia*, which includes about 132 species of metallic, medium-sized, plant-feeding scarabs. The Japanese beetle is *P. japonica*; it was accidently introduced to the United States early in the last century, and it has become one of the most significant plant pests in eastern North America.

This hierarchal approach to classification is roughly analogous to our system of assigning addresses to people (Fig. 2.6). The kingdom is roughly analogous to the continent you might live on; the phylum, your country in the continent; the class, the state you live in; the family, your city; the genus, your street; and the species, your house number. In a very real sense, the scientific name of a species is its address in the great, beautiful, rich, and fascinating sweep of living things that inhabit this little blue ball called Earth.

The scientist who "discovers" and describes a new species gets to decide what its scientific name will be—within certain rules. The genus, if the new species belongs to an existing genus, is fixed; a new species of *Popillia* must have *Popillia* as its generic name. However, the specific epithet can be pretty much anything, as long as it is unique within the genus and isn't overtly vulgar. With almost a million species described so far, some systematists have gotten creative with the names they've assigned. Look at the list in Table 2.1 and say these names out loud…

| Table 2.1 | *Some interesting scientific names of insects* |
| --- | --- |
| *Dicrotendipes thanatogratis* | A small fly named for the Grateful Dead |
| *Cryptocercus garciai* | wood roach |
| *Eubetia bigaulae* | tortricid moth |
| *Pieza rhea* | mythicomyiid fly |
| *Verae peculya* | braconid wasp |
| *Leonardo davincii* | beautiful moth |
| *Reissa roni* | a microbombyliid fly |
| *Polemistus Chewbacca, P. vaderi* | "Star Wars" wasps |

*continued*

**Table 2.1** Some interesting scientific names of insects (continued)

| | |
|---|---|
| *Heerz lukenatcha* | a small parasitic wasp |
| *Lalapa lusa* | a tiphiid wasp |
| *Enema pan* | a rhinoceros beetle |
| *Preseucoela imallshookupis* | A tiny wasp |
| *Pheidole harrisonfordi,* | an ant |
| *Agra katewinsletae, A. schwarzeneggeri* | carabid beetles |
| *Myrmekiaphila neilyoungi* | trapdoor spider |
| *Aptostichus stephencolberti* | trapdoor spider |
| *Agathidium bushi, A. cheneyi* | slime mold beetles |

## Some Important Classes of Arthropods

We will be discussing the diversity within the Hexapods in a later chapter, but now I'd like to introduce you to the other classes of arthropods, since, while they are perhaps not as overwhelmingly diverse as insects, they are critically important to our understanding of evolution and ecology, and many are medically or economically important. Being able to recognize these classes will enrich your understanding of the natural world.

**Xiphosura**: The horseshoe crabs (Fig. 2.7) are the ancient, marine relatives of spiders and other archnids we mentioned earlier. Together with the arachnids, the Xiphosura belong to a larger group (a "superclass") called the Chelicerata; all Chelicerata have more or less "fang-like," often venomous, mouthparts. The Xiphosura is a very small class—there are only four species of horseshoe crabs in the world. All have a hard, shield-shaped, chitinous carapace concealing the bulk of the body and their legs, and a long, fearsome-looking but harmless tail called a telson, which they use to right themselves should they be flipped on their backs. They have five pairs of legs, breath through "book gills," and prey on seafloor invertebrates.

**Arachnida:** The spiders, scorpions, mites, ticks and others (Fig. 2.8) are extremely diverse, with about 80,000 described species and, undoubtedly, thousands of undescribed species. Arachnids typically have eight legs (although the immatures of some only have six), bodies organized into two main regions, and those chelicerate mouthparts; they also lack antennae. Beyond this they assume a tremendous range of forms. Almost all Arachnids are either predators or parasites of other animals, often other arthropods (although there are some plant feeding mites), and they feed on the liquid, or liquefied, portions of their prey/ hosts. Some arachnids are important biological control agents for pests; others are medically important either because of their venomous bites or stings (some spiders and scorpions) or their ability to transmit disease (ticks).

**Figure 2.7.** Horseshoe crab.

14    Insects and People

**Figure 2.8a–d** A selection of Arachnids.

**Crustacea:** The crustaceans (Fig. 2.9) are considered by most modern systematists to be a superclass containing six classes; the most familiar Crustacea are the Malocostraca—the lobsters, shrimp, crabs, and their relatives. The 50,000 or so species of crustaceans are highly variable, but all have mandibulate ("chewing") mouthparts and, uniquely, two pairs of antennae. Their appendages are *biramous*, meaning there are two parts arising from the same point of origin. Most incorporate calcium into their chitinous exoskeletons, making them harder and more rigid. Almost all are marine or freshwater animals, and many are extremely valuable (and tasty!), as anyone who has seen *Deadliest Catch*, or dined at a Calabash seafood house, can testify. Crustaceans dominate marine ecosystems in much the same way as insects dominate terrestrial ecosystems; indeed, their dominance in the ocean may explain in part why there are so few insects in the sea. They fill virtually all ecological roles animals might fill. They are predators, grazers, detritus feeders, and parasites; some

**Figure 2.9a–b** A selection of Crustaceans.

The Family Tree: Sorting Out a Million Critters 15

bore through coral rock, and others bore through ship's timbers. One species, the Antarctic krill (Fig. 2.10), is probably the single most abundant animal species on the planet; they feed countless seals, whales, fish, and other animals. The heaviest living arthropod is a crustacean (the Maine lobster, Fig. 2.11); the largest, the Japanese spider crab (Fig. 2.12), which has legs that span as much as 12 feet.

**Chilopoda:** The centipedes (Fig. 2.13) are generally long-bodied animals with one pair of legs per each of their many segments (*centipede* means "100-feet," although none actually have 100 legs, but some have up to 80!) Their bodies are dorso-ventrally (top to bottom) flattened, and they have one pair of well-developed antennae and generally poorly developed eyes. All are active, terrestrial predators on other animals, and the legs on their first segment are modified into venomous "fangs" that they use to capture and manipulate prey. There are about 3,000 or so described species, with the greatest diversity in the tropics. Some of the larger species (and the biggest is about 12 inches long) can deliver a painful, and in some cases, dangerous bite.

**Diplopoda:** The Millipedes (Fig. 2.14) are generally elongate, slow-moving, many-segmented creatures with two pairs of legs per segment. *Millipede* means "thousand feet," but no millipede actually has that many, although one species does have about 700! Millipedes have chewing, mandibulate mouthparts, and all feed on plant matter, detritus, or fungi. They are generally round in cross section, have one pair of well developed, but usually short, antennae, and, like crustaceans, many have exoskeletons reinforced with calcium. The largest millipedes are about 14 inches long, although most species are much smaller. While none is venomous, they do have some interesting defensive mechanisms: many roll into a tight ball, protecting their vulnerable undersides, when threatened, and some have projections along their sides that may make it difficult for smaller predators to handle them and ingest them. Most significantly, many exploit "chemical warfare" to defend themselves. Some smell like almonds or cherries when disturbed—these are producing cyanide as a defensive chemical, while others produce caustic and foul smelling quinones or terpenes. Millipedes are ecologically important in their role as primary consumers and decomposers, and many are prey for other animals; rarely, they are agricultural pests.

**Figure 2.10.** Antarctic krill.

**Figure 2.11.** Maine lobster.

**Figure 2.12.** Japanese spider crab.

**Figure 2.13a–c** A selection of centipedes from the class Chilopoda.

**Figure 2.14a–b** A selection of millipedes from the class Diplopoda.

# Where Are the Bones?
## The Insect Exoskeleton

Insects, like all other animals, must deal with assaults of all kinds as they make their way through the world. There are things trying to eat them, there are things trying to infect them, there are things that might puncture them or otherwise inflict injury. As typically little-bitty, terrestrial animals, living in a great big, and, by comparison, dry world, most also have to jealously guard the water their bodies contain. One of the keys to the enduring success of the insects, and for that matter, the other arthropods, is the support structure for their bodies: their exoskeleton.

The insect exoskeleton functions as a scaffold for supporting the body, a framework for the attachment of muscles for locomotion, a suit of armor that protects the insect from all manner of assaults from the outside, and an almost watertight life suit (Fig. 3.1). In the size scale of terrestrial insects, it is an extremely strong yet light-weight solution to supporting a body, and because of how it grows and is built, it can be modified to solve a huge range of life-support problems. But in spite of all the benefits of an exoskeleton, there are also significant limitations to this sort of body architecture. In this chapter, we'll learn about how the insect exoskeleton is put together, and we'll also look at the advantages and disadvantages of an exoskeleton.

**Figure 3.1.** An insect in its exoskeleton.

## Chitin: One of Nature's Most Versatile "Plastics"

The insect exoskeleton is so functional and versatile largely due to the materials that it is made of. One of the most prominent components is a chemical called "chitin." Chitin is a polymer (a large molecule made of linked, repeating units of a smaller molecule) of a special sugar called n-acetylglucosamine, which is essentially a glucose molecule with a nitrogen-containing amine replacing one of the hydroxy groups (Fig. 3.2). Chitin is found in many other living things. It's found in the cell walls of fungi, the exoskeletons of roundworms, and the beaks of squid and the "teeth" of snails. The repeating sugar molecules form long chains; the chains bind to each other to make microfibers. Chitin itself is a tough, flexible material that is resistant to water, mild acids, and mildly caustic substances. In insect exoskeletons, chitin is always embedded in a matrix of protein, making what we call a composite material. The insect exoskeleton is composed of sheets of chitin in this matrix. Often, adjoining sheets have their microfibers lined up in differing orientations, greatly increasing the strength and puncture resistance of the exoskeleton. Like many synthetic plastic materials, the chitin-protein composite can be "molded" into a huge range of shapes while maintaining its strength and structure and, as we'll see in a bit, insects exploit this property abundantly.

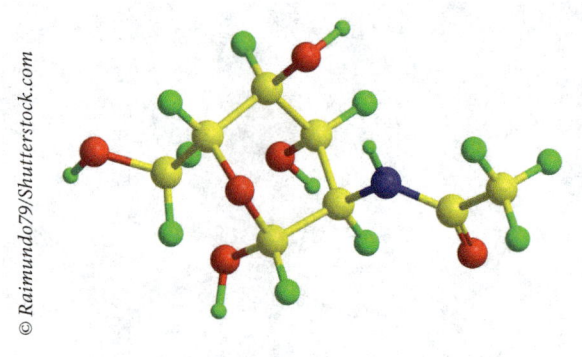

**Figure 3.2.** The chemical structure of n-acetylglucosamine, the building block of chitin.

Chitin is chemically very similar to another important biological polymer—cellulose. Cellulose is composed of chains of glucose—the same sugar modified to acetylglucosamine in chitin. Cellulose is the primary structural element in the cell walls of plants and algae; it is the most abundant biogenic polymer on earth. Wood is composed primarily of cellulose and another group of polymers, lignins. After these two chemicals, chitin is the most abundant biologically derived polymer, and certainly the most abundant animal-derived polymer.

Chitin, and a product made from it, called chitosan, have a huge range of industrial and commercial applications. They can be used for clarifying wine and beer, used to make self-healing paints, used in waste-water treatment, used as anti-fungal seed treatments, and used to make filters. Importantly, there are a number of very significant medical applications—chitosan is a life-saving antihemorrhagic that can stop severe bleeding very quickly, and both chitin and chitosan can be used to make dressings that speed the healing of severe burns and other grave wounds. Chitosan also shows some promise in reducing the growth of cancerous tumors and may have some utility in weight reduction. Most commercially produced chitin and chitosan are extracted from crab or shrimp shells, or from fungi.

## The Insect Exoskeleton Up Close

In order to understand all that an exoskeleton can do, we first need to understand how it is put together, so we'll briefly take a look at the anatomy of the exoskeleton (Fig. 3.3). If we start from the inside out, the first structure we encounter is the **basement membrane**. The basement membrane is a non-living structure produced by the living epidermis immediately above it; it is essentially a thin sheet of chitinous material. The purpose of the basement membrane is to provide a gate to the living epidermis from all the things going on inside the insect's body, and it provides a substrate to support the epidermis. It is composed of sugar polymers and other substances and is made, at least in part, by the epidermis.

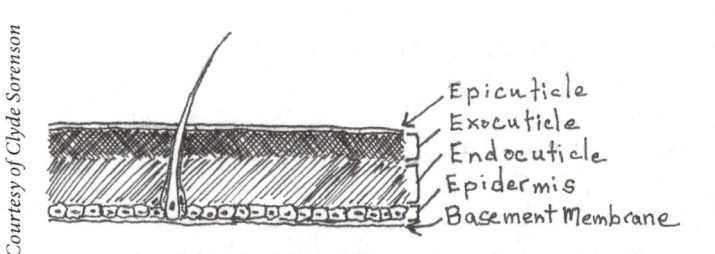

**Figure 3.3.** Cross-section of insect exoskeleton.

Immediately above the basement membrane is the **epidermis**. This is the only living part of the exoskeleton, and its primary function is to build the rest of the structure. In most places, the epidermis is but a single cell layer thick, and the entire exoskeleton (everything you see when you look at the insect) and some internal structures, are lined with this epidermal layer.

20  Insects and People

The epidermal cells secrete the rest of the exoskeleton from compounds made by other cells in the body in a precise process we'll discuss later.

Above the epidermis is the **cuticle**, which can have as many as three layers. The inner two layers contain chitin, proteins, and sometimes other chemicals. The innermost of these is the **endocuticle**. Endocuticle is composed of as much as 50% chitin in a protein matrix (making that composite material we talked about earlier), and it is tough, but flexible. Endocuticle is most apparent in areas on the insect's body where flexibility is needed; most of a caterpillar's body, for instance, is clothed in a primarily endocuticle type of exoskeleton.

Above the endocuticle, in some parts of the exoskeleton, is the **exocuticle**. Exocuticle has a lower proportion of chitin and a higher proportion of protein; the proteins found in exocuticle are somewhat different from those found in areas dominated by endocuticle. Addtionally, exocuticle contains chemicals called quinones, which form strong cross-links between the protein molecules, making the exocuticle more rigid and hard. The process that forms exocuticle is called tanning or sclerotization, and another name for exocuticle is sclerotin. Exocuticle is most highly developed in places on the insect's body where protection, strength, or durability is necessary. On a caterpillar's body, the most highly developed exocuticle is found in the insect's mandibles or jaws, which have to cut and chew the many leaves the caterpillar will eat as it grows. Much of the body of most beetles has a very well-developed exocuticle layer, which protects these animals as they make their way through the often abrasive habitats they live in.

The last, outermost, and thinnest layer of the cuticle is in many ways the most important. This is the **epicuticle**, and unlike the endocuticle and exocuticle, it contains no chitin. The primary function of the epicuticle is water-proofing, but primarily, to keep water in! Very small, terrestrial animals, such as most insects, have a very high surface to volume ratio; an important consequence of this is that these animals can lose the relatively small volume of precious water their bodies contain very, very rapidly, through evaporation from the proportionally very large surface area. In insects, the epicuticle prevents this. The epicuticle is composed of multiple layers of waxes and cements that cover the entire exoskeleton.

The critical water-conserving roll of the epicuticle is very easily demonstrated when it is compromised in some way. If the epicuticle is abraded by a material such as diatomaceous earth, or washed off with a detergent of the right sort, the affected insects will die extremely rapidly. I used this to good effect when I lived in Reno, NV, where every fall my house was invaded by huge numbers of boxelder bugs (Fig. 3.4) seeking an overwintering site. I would spray aggregations of hundreds of the insects with a solution of insecticidal soap; in the dry air of the desert, they would start dying before the spray solution itself would dry on them!

Insects also incorporate other substances into their exoskeletons in some cases. The mandibles of some insects, especially some caterpillars and beetles, are hardened by the inclusion of zinc or manganese. Bands or pads of an extraordinarily rubbery protein called resilin are often found in the joints of the wings and legs of many insects. Resilin is actually more elastic than rubber

**Figure 3.4.** Boxelder bugs, sucking insects from the order Hemiptera that feed on the seeds of maple trees.

Where Are the Bones? The Insect Exoskeleton

itself and is one of the most elastic natural substances known. Resilin in insect joints stores potential energy when stretched or compressed, assisting muscles and increasing the efficiency of movement. Resilin also replaces muscles in some places in some insects, saving weight and metabolic energy.

The composite material that makes up insect cuticle can take a huge variety of shapes and forms, and many structures we find on exoskeletons can help us identify who it is we are looking at and understand how they live. In many places on most insect exoskeletons, there are invaginations that look like little valleys in the exoskeleton (Fig. 3.5); these are often **sutures** that form points of attachment for muscles. In other places, there may be small **pits** (Fig. 3.6). Some of these are sensory, while others may again be muscle attachments. Insect exoskeletons can be "decorated" with large, moveable, defensive **spurs** (Fig. 3.7), or large, immovable spines (Fig. 3.8). Some insects are notably "hairy" (Fig. 3.9); while these fuzzy structures look like hair, these structures develop in a completely different way from mammalian hair and are more properly called **setae** (singular *seta*). Setae can perform defensive function, as in the caterpillar (Fig. 3.10), or a sensory function. Some defensive setae and spines are armed with venom glands. Some other insects, notably the Lepidoptera (the moths and butterflies), have wings covered in flattened, scale-like setae which contain the pigments or structures that give the wings their colors (Fig. 3.11). Contrary to popular myth, rubbing the scales off the wings of a butterfly won't disable its ability to fly! Many insects have, at least by our estimation, outrageous outgrowths of different parts of the exoskeleton that have a variety of purposes (Fig. 3.12). As we'll see later, insects do indeed exploit the versatility of a chitinous exoskeleton in a tremendous variety of often quite astounding ways.

**Figure 3.5** Grasshopper head and thorax, showing numerous sutures.

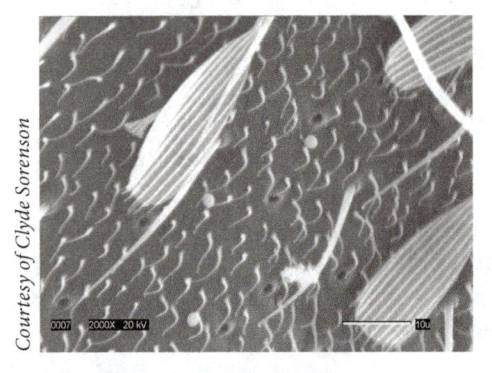

**Figure 3.6.** SEM of genitalia of a male mosquito showing sensory setae and tiny sensory pits.

**Figure 3.7.** Tibial spurs on the leg of a mole cricket.

**Figure 3.8.** Cecropia moth caterpillar with many immovable spines decorating its exoskeleton.

22  Insects and People

**Figure 3.9.** Bumblebee with dense covering of hair-like setae.

**Figure 3.10.** A banded tussock moth caterpillar with dense setae covering its exoskeleton.

**Figure 3.11.** Close-up of wing of lunate zale moth, showing rows of colored scales.

**Figure 3.12.** Two Hemipterans with extravagant outgrowths of their exoskeletons.

## Advantages of an Exoskeleton

1. Strength to weight: An insect exoskeleton is, essentially, a series of connected tubes of various shapes and lengths, and this is one of the keys to its utility. A tube constructed out of a given amount of material will typically be much stiffer than a solid rod made of the same amount of material. The tube has a higher strength-to-weight ratio. At the scale of most living insects, a skeleton made of external tubes is going to be the strongest and most "economical" support structure in weight.

2. Waterproofing: We've already discussed this advantage, but it bears repeating: the epicuticle helps the insect guard the small amount of precious water its body contains.

3. Efficient muscle attachment: Insect muscles, like all muscles, can only do work by pulling. Insect muscles can pull very efficiently inside the series of tubes that is the insect's body.
4. Protection from infective agents: The cuticle is an excellent barrier to many (but not all!) micro-organisms that might exploit the insect.
5. Protection against predators: Again, the cuticle can provide an excellent barrier to many (but not all) predators.

Can you think of others?

## Disadvantages of an Exoskeleton

1. Lack of flexibility: A suit of armor made of rigid tubes restricts movement to some, and sometimes great, extent.
2. An exoskeleton limits size for terrestrial animals: While an exoskeleton provides excellent strength to weight at small scales, at some point, in order to support the weight of the animal, the walls of the exoskeleton's tubes would have to become too thick to contain muscles large enough to move the body parts around. Living insects clearly don't approach this limit, since ancient insects grew much larger, but this scaling factor would prevent insects reaching the size of many mammals.
3. In order to grow, an animal with an exoskeleton must **molt**. Exoskeletons do not stretch, and the only way to get a bigger one is to grow it inside the old, and then molt. We'll be discussing the molting process in a subsequent chapter, but for now, it's important to understand that making a new exoskeleton is an expensive process in metabolic terms, and insects actively engaged in molting are extremely vulnerable. Their new cuticle is soft until it tans and offers poor support, so newly molted animals can't escape as readily as those with a fully tanned cuticle. (Many insects molt at night and in secluded places, if possible, to limit the chances of getting eaten while molting.) This presents another limitation to the size of terrestrial arthropods in another very important way, in that terrestrial arthropods are subject to the demands of gravity. Very large, terrestrial arthropods could have great difficulty in maintaining the form of their body and supporting their body weight before the new cuticle tanned. Aquatic arthropods face no such limitation since their bodies weigh essentially nothing in the water; this is why the largest arthropods are the marine crustaceans we discussed earlier.

Exoskeletons clearly contribute to the success of the insects. Protected by their durable cuticles, they can live in virtually any habitat and confront virtually any assault. But there are many other reasons insects are so dominant, and we'll be addressing some of these in subsequent chapters.

# IV

# Eyes, Ears, Nose, and Throat: The External Anatomy of the Insect Head

One of the many things about insects that creep some people out over them is their rather alien body architecture. Their segmented bodies, long antennae (in some insects), and odd, faceted eyes all serve to distance them from most people. But in actuality, these very traits contribute to the success of insects. Insect bodies are compartmentalized; each of the three major body regions houses different, important functions. In our examination of the external anatomy of the insect body, we shall start at the beginning—with the insect head.

The insect head is responsible for three important life functions: containing the brain, taking in food, and sensing the outside world. Most of the obvious structures we see on the head have something to do with the latter two of these jobs. As we begin our examination of the head, it is important to understand that while it looks like a single structure, ancestrally it is derived from the fusion of six primitive segments, each with a pair of appendages (Fig 4.1). The mouthparts we see on insect heads, and, apparently, the antennae, are derived from the appendages on those six ancestral segments. These same appendages, on that ancestral beast that eventually evolved into the modern insect, were used for locomotion, so some of these bits look very leggy indeed, but more on that later. For now, let's tackle the sensory organs we find on the insect head.

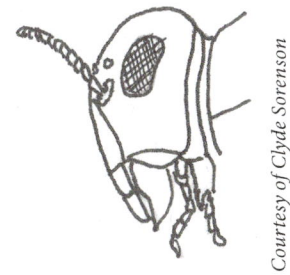

**Figure 4.1** Diagram of putative origin of the insect head.

## Insect Eyes

Many insects actually have two kinds of eyes—the large, obvious, compound eyes, and much smaller, "simple" eyes called ocelli or stemmata. We'll discuss the compound eyes first.

Compound eyes are found only on adult insects and on the immatures of insects that undergo incomplete, or simple, metamorphosis. If we look at the structure of a compound eye, we can see what the term "compound" is referring to: the eye is made up of many smaller units, each with its own lens forming a facet on the surface of the eye (Fig. 4.2). Each of these individual units is called an **ommatidium** (plural **ommatidia**). The number of ommatidia in an insect's eye can vary wildly;

**Figure 4.2.** Close-up of fly compound eye, showing the lenses of individual ommatidia.

**Figure 4.3.** The compound eyes of the harlequin darner dragonfly, a vigorous aerial predator of other insects.

some insects have only a few in each eye, while others, particularly active predators, can have almost 30,000 in each eye (Fig. 4.3).

There is a great deal of misunderstanding about how and what insects see. Since each eye is composed of many subunits, each with its own lens, many folks think they must see multiple copies of whatever they look at (Fig. 4.4). However, if one gives this a little thought, it is quickly apparent that this can't be what insects see—how in the world would they know which image to respond to! Our best understanding suggests that, in most insects, each ommatidium captures a color average of the scene it is aimed at and sends to the brain just that information. The brain then integrates all these dots of color to form a mosaic image—in other words, each ommatidium sends one "pixel" of information to the brain, and the resulting image is a composite of all the "pixels," much as a newspaper picture or image on a television screen is composed of dots of pure color. (Get really close to your TV sometime, maybe when you have nothing better to do, and you'll see the tiny dots that form the image.) The "quality" or resolution of the image thus formed is dependent on the number of ommatidia in each eye contributing "pixels" to the image. In general, the more ommatidia, the better the resolution. Some insects, with extremely large eyes and tens of thousands of ommatidia, can see very well indeed, but they don't have anything like the resolving capability our eyes have, and the best insect vision is probably myopic and fuzzy compared to ours.

We should mention that the eyes of some insects, particularly night-flying moths and beetles, don't work quite the same way. In these insects, the multiple facets supply light to multiple points on a single, continuous retina located deep in the eye, and form a single image. These eyes make the most of dim light, but they may not provide the same level of resolving power more conventional eyes do.

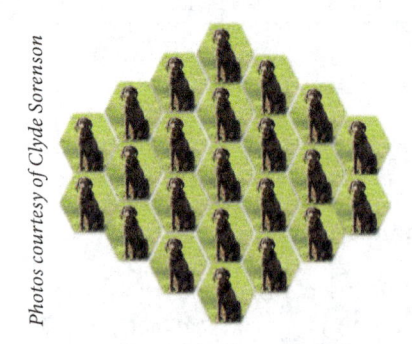

**Figure 4.4a.** An erroneous idea of what an insect might see when looking at my puppy, Hatteras.

**Figure 4.4b.** What an insect probably does see.

**Figure 4.4c.** What we see.

26   Insects and People

Insects can see color, although the colors they can see often differ from those we see. Insects can see greens, yellows, blues, violet, and even ultraviolet (which we can't see), but most can't see red. To these insects, red will look black. Due to the construction of their eyes and some unusual characteristics of their nervous systems, insects also detect motion extraordinarily well. A good measure of this ability to detect movement is the flicker-fusion rate—the rate at which individual flashes (or pictures) cease to be detected as individuals and merge into a single, steady light (or motion picture). Some insects have a flicker fusion rate more than five times higher than that for humans; for these insects, watching a movie is actually more like watching a stream of still pictures!

**Figure 4.5.** A stonefly, showing the three ocelli between the compound eyes.

Structurally, ocelli and stemmata are somewhat similar to a single ommatidium, but they are functionally different. **Ocelli** are found on adult insects and on the immatures of insects that undergo simple metamorphosis. There are usually three, located in a triangle between the eyes (Fig. 4.5). They don't appear to be able to form resolved images, but they are quite sensitive to low levels of light. Their exact function is not well understood, but in some insects, they seem to be used to keep track of the horizon as the critter flies, while in others they appear to have a role in setting the circadian clocks they use to keep track of the time of day and season of the year.

**Stemmata** are somewhat different in structure from ocelli and communicate with a different part of the brain. They are most common on the larvae of insects that undergo complete metamorphosis and are most well known on caterpillars (Fig. 4.6). Stemmata, again, don't appear to be able to resolve images, but they can detect colors and movement, and they are important in the defensive reactions of these insects.

**Figure 4.6.** The head of a silver spotted skipper caterpillar; the five raised bumps along the edge of the yellow eyespot are ocelli-like stemmata. The yellow eyespot is simply a colored spot on the exoskeleton and plays no role in vision.

# Antennae

The other major sensory organs on the heads of most insects are the antennae. Many folks call these often long, wavy structures "feelers," because some insects do appear to touch things with them. In most insects, however, the antennae are the seat of the sense of smell, or **olfaction**. They perform the same sensory function as our noses do. An insect antenna has three primary segments: the stipe, pedicel, and flagellum (Fig. 4.7). All the muscles that move the antenna about are located in the head capsule and basal two segments; no muscles are typically found in the flagellum. The flagellum is where most of the sensory power of the structure is located. Basically, it is a hollow, fluid-filled tube with nerves ending in olfactory sensors. The sensors are typically located at the bottoms of tiny pits which capture the scent molecules from the air (Fig. 4.8). Most insects have olfactory systems tuned to a few critical stimuli, including food, water, and potential mates.

Most of the fancy variations in the shape and size of insect antennae (Fig. 4.9) enhance the ability of different insects to smell pertinent odors in different environments. The ancestral antennal form has a longish flagellum that appears to be segmented (Fig. 4.9a). Many beetles have clubbed antennae, often with lobes that can be expanded (Fig. 4.9b and d); some moths have huge, feather-shaped antennae (Fig. 4.9g). Most of these modifications serve to increase the surface area of the antennae and therefore, the sensitivity. The males of some moths, notably the giant silkworms such as the luna moth, can detect the scent of a receptive female from miles away, thanks to the huge surface area of their antennae. The fact that many insects have two, relatively long antennae allows the creatures to directionalize the source of odors. If one of the antennae is catching a particular scent while the other is not, that antenna is obviously closer to the source, and the insect will move accordingly.

While olfaction is the primary sensory purpose of the antennae in most insects, in some they perform other functions as well. In many higher flies, they function in part as a wind or flight speed indicator; these antennae have a large "spine" that deflects more as the insect travels faster (Fig. 4.9c). In one exceptional beetle, they've assumed a defensive role—the tips are extremely sharp and armed with a venom gland!

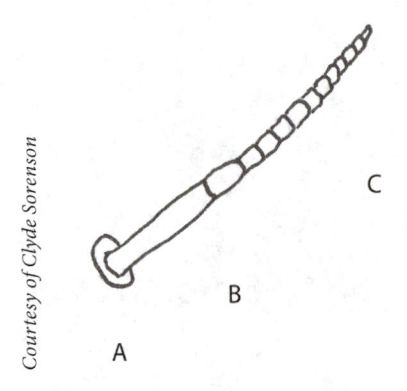

**Figure 4.7.** Parts of an insect antenna.

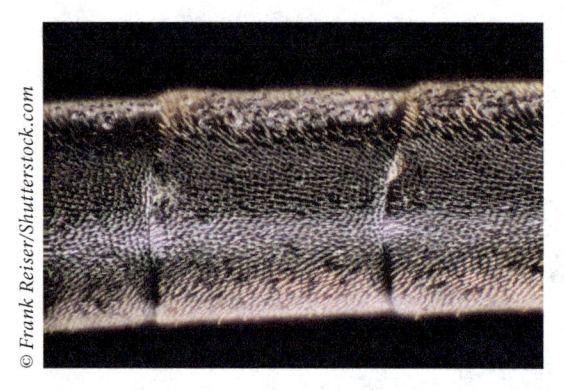

**Figure 4.8.** SEM of a sensory pit.

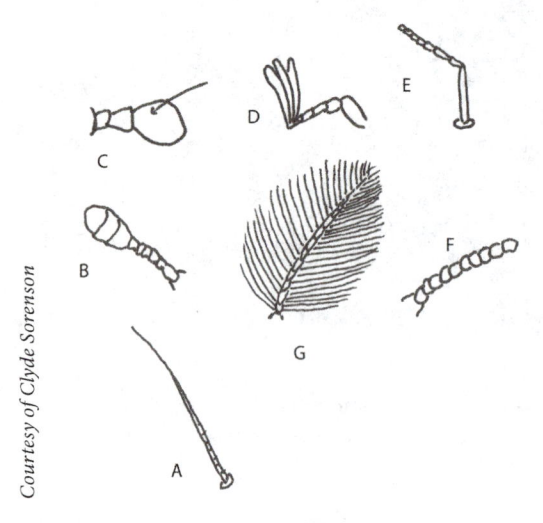

**Figure 4.9.** A selection of insect antennae.

# The Insect Mouth

All insects, at some point in their lives, must eat, just as all other animals do, and the kinds of mouthparts insects have are often one of the defining characteristics we use to describe the major groups or orders of these animals. As I suggested earlier, the mouthparts of modern insects evolved from ancestral walking appendages, although some of the modern parts don't look much like legs any more. More primitive insects have chewing, or mandibulate, mouthparts; all other types of mouthparts evolved from these. In order to understand how insects feed, and to also understand how other kinds of insect mouths evolved, we need to understand how these chewing mouths are organized.

We'll start our examination of the insect mouth from the front, and work our way back. The first structure we come to is the **labrum** (Fig. 4.10a), which essentially functions as the top "lip" of the critter's mouth. While the labrum is a single structure, it is actually two of those ancestral legs fused together. In many insects that feed on plants, the labrum is notched, to help the insect keep its mandibles in line with the edge of the leaf it is eating.

28   Insects and People

Behind the labrum are the **mandibles** (Fig. 4.10b), the actual, working jaws of the insect's mouth. These are a bit more leg-like in appearance, and, in most insects, are the hardest, most durable structures in the entire body. They are typically serrated in one way or another to facilitate the cutting of the tissue the insect is eating. An insect's mandibles work in a plane perpendicular to the long axis of the animal's body; our jaws work more or less parallel to the long axis of our bodies.

Behind the mandibles are the **maxillae** (Fig. 4.10d), which are probably the "leggiest" looking structures in the mouths of most insects. The maxillae (singular maxilla) are used to manipulate the food, and the main body of the structure often has a hardened hook at the tip to facilitate this; the leggy-looking bit is the **palp**, which often is loaded with taste receptors.

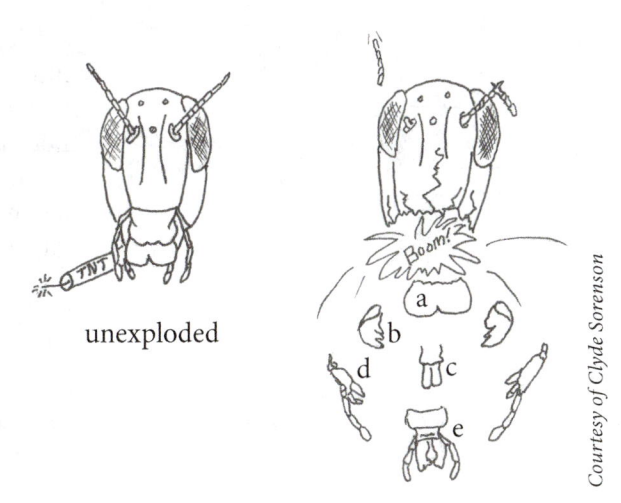

**Figure 4.10.** "Exploded" diagram of basic insect mouthparts. The parts are: a. labrum; b. mandible; c. hypopharynx; d. maxilla; e. labium.

Behind the maxillae is the **labium** (Fig. 4.10e), a fusion of two ancestral appendages much like the labrum. The labium functions as the "bottom lip" of the insect mouth, preventing food from falling out the back of the insect's mouth as it eats. The labium typically has a pair of sensory palps much like those on the maxillae.

Tucked inside the mouth is the **hypopharynx** (Fig. 4.10c), which functions somewhat like a tongue to manipulate food within the oral cavity in insects that eat solid food.

Even within those insects that eat solid food with mandibulate mouthparts, there are tremendous variations in how those mouthparts look. Many predaceous insects have mandibles with large, pointed cusps near the tips and shearing cusps towards the back. Mandibles also vary among insects that eat plants; those that eat tough, fibrous plants have stout, crushing cusps. In some predaceous beetles, the mandibles are long, sickle-shaped tools with a groove along the inner surface. They use these jaws to pierce their prey, and then draw the liquid from its body through the grooves. Probably one of the most remarkable variations on the basic chewing mouth organization occurs in the aquatic immatures of dragonflies and damselflies (order Odonata); in these insects, the labium has evolved into a large, muscular, articulated "mask" that covers the rest of the mouthparts (Fig. 4.11a). The dragonfly nymph uses this mask to capture its prey. The labium lashes out, and the toothed palps grab the prey (Fig. 4.11b and c). The snake-like strike of a dragonfly nymph is one of the fastest movements in the animal world! In some other insects, some of the mouthparts may be co-opted for other purposes; in male stag beetles, for example, the mandibles have become antler-like structures they use in competitions over females (Fig. 4.12).

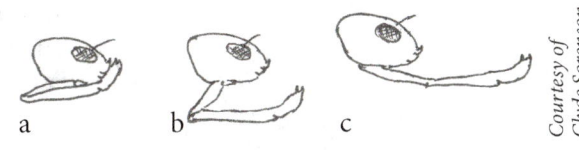

**Figure 4.11.** Extension of the labial mask of a dragonfly nymph.

**Figure 4.12.** The antler-like mandibles of a male stag beetle.

Eyes, Ears, Nose, and Throat: The External Anatomy of the Insect Head

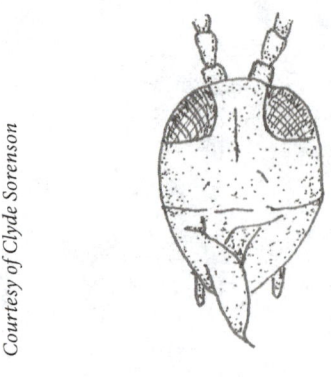

**Figure 4.13.** Poorly drawn (by me) representation of thrips mouthparts.

**Figure 4.14.** Bee mouthparts.

**Figure 4.15.** Piercing-sucking mouthparts of the wheel bug, a large, predaceous assassin bug of the order Hemiptera.

While mandibulate mouths are very common, there are a number of variations in insects that feed on non-solid foods in many different ways. One of the more ancient of these modifications is found in the thrips (order Thysanoptera), tiny insects that feed on a range of substrates. They have only one mandible, which they use to tear holes in the tissue upon which they are feeding, and then use a tube constructed of parts of their maxillae to suck up the released liquid (Fig. 4.13). As they feed, they repeatedly "spit" digestive enzymes into the wound to release the nutrients contained in the tissue. While most feed on fungi, some are important plant pests, and some are important vectors of plant diseases others are predators on other insects, primarily other thrips.

Many bees have all-purpose mouths that are efficient at handling both solid and liquid substances (Fig. 4.14). Their mandibles are well adapted to manipulating the pollen they use to make the "beebread" they feed their young. The bee's labium is modified to function as an effective "tongue" for lapping up the nectar they use to make honey.

A great range of insects has mouths adapted to **piercing** something and then **sucking** the fluids out of it, much like a hypodermic needle can be used to draw blood from your arm. Before we get into any details, it is important to remember that, no matter how bizarrely different the mouthparts of these insect look, compared to mandibulate mouths, they are built of the same pieces and parts, although they may be highly modified.

In the order Hemiptera (the true bugs, aphids, cicadas, and such), the "beak" is composed of **stylets** which are highly modified parts of the maxillae and mandibles (Fig. 4.15). These fit into a groove in an extended labium, and they lock together to form two tubes: a salivary channel, though which they inject digestive enzymes and/or venom, and a larger feeding tube through which they suck up their partially digested, liquid meals.

In the biting Diptera, or true flies (mosquitoes, etc.), the feeding stylets are composed of a highly modified labrum and hypopharynx, again forming a salivary tube and a feeding tube (Fig. 4.16). The maxillary stylets are the bits that actually pierce your skin. These parts are all folded into a protective sheath made by the labium. When a mosquito bites, it actually "saws" its stylets into your skin by vertically pumping then up and down along-side each other. Horseflies have knife-like, serrated mandibles that they use like wicked little scissors, to quickly flay open skin (Fig. 4.17); they suck the released blood through a channel on the back of the labrum. Other flies have sponging mouthparts (Fig. 4.18) that blot up liquid substances. Most of these flies regurgitate digestive enzymes onto potential food sources and can transfer disease agents in the process.

**Figure 4.16.** Piercing-sucking mouthparts of a mosquito.

**Figure 4.17.** Horsefly mouthparts.

While the caterpillars of Lepidoptera (moths and butterflies) feed on solid foliage with chewing, mandibulate mouthparts, most adults have **siphoning mouthparts** adapted for feeding on nectar and other liquids (Fig. 4.19). Their siphoning "tongues" are actually highly modified maxillary lobes, one from the left and one from the right, hooked together for most of their length. In most Lepidoptera, the proboscis is coiled up under the head when the insect is not feeding; the moth or butterfly can extend the structure into a flower (or puddle of puppy piddle) through hydraulic pressure, and then simply suck the liquid into its gut through the paired feeding tubes. Tiny muscles help coil it back up on the way to the next flower. This allows some moths to feed from flowers with really deep corollas, like petunias and tobacco flowers; one moth has a proboscis approaching 10 inches, more than four times its body length! (We'll come back to this moth a bit later.)

In some adult insects, notably the Ephemeroptera (mayflies) and many moths, the sole responsibility of the adult stage is to reproduce; in many of these insects, the mouthparts are vestigial, meaning they are greatly reduced and non-functioning. In these insects, all the energy the adult will need to successfully disperse, mate, and lay eggs is harvested by the immature stages and stored, as fat, in the body.

**Figure 4.18.** Sponging mouthparts of a true fly (Diptera).

**Figure 4.19.** Siphoning mouthparts of a Lepidopteran.

It's important to remember that all the structures we've talked about in this unit are parts of the exoskeleton and the surface of all are shed and replaced along with the rest of the exoskeleton when the insect molts.

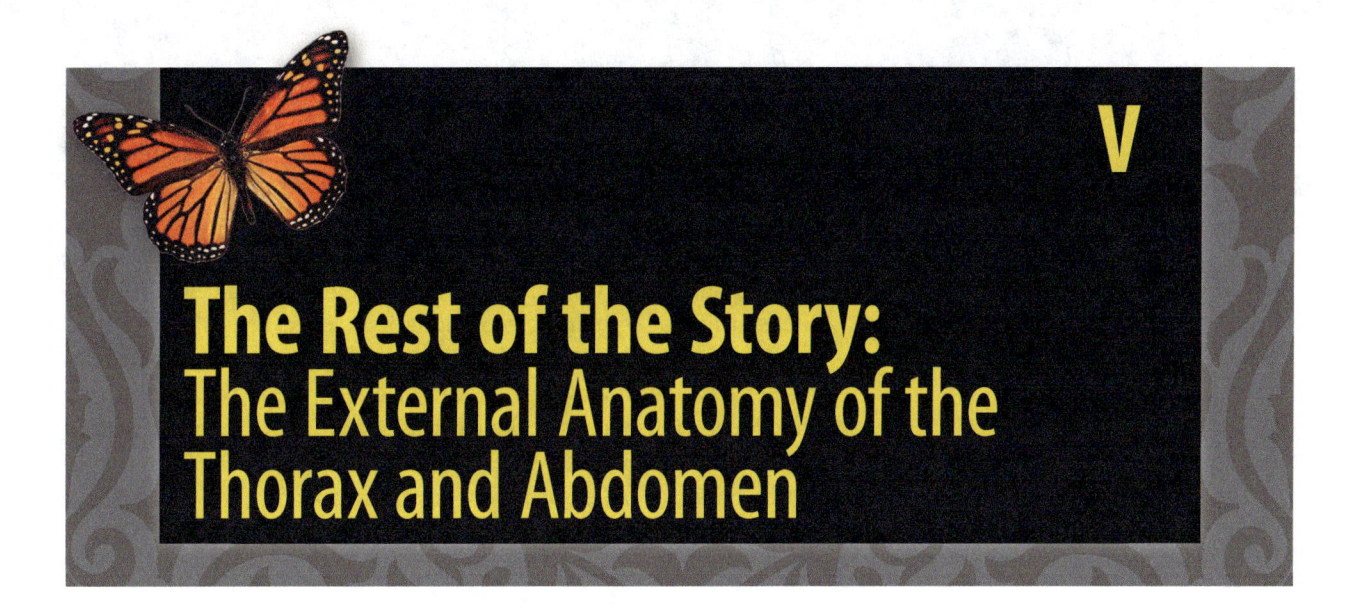

# The Rest of the Story: The External Anatomy of the Thorax and Abdomen

The insect's body, astern of the head, with its six legs, wings, and odd little twiddly bits hanging off the back end, is just as alien to many folks as the head we just discussed in the previous chapter. However, in the insect body plan, evolution has produced an efficient and robust form. The compartmentalization of function continues in the remaining two regions of the body: the thorax is responsible for locomotion, while the abdomen houses most of the life functions of the creature. In this chapter, we'll take a look at the external morphology of these body regions.

## The Thorax

The insect thorax (Fig. 5.1) is the center of locomotion in the insect; it supports all the structures that allow the animal to get around in the world. The thorax is composed of three ancestral segments, never more, never less. The first segment, immediately behind the head, is the **prothorax**; the middle, the **mesothorax**; and the last, nearest the abdomen, the **metathorax**. Each segment is composed, more or less, of four plates, or **sclerites**, which often are marked by sutures that provide muscle attachment and/or rigidity. Together, these plates more or less take the form of a four-sided box. The top sclerite is the **notum**; the bottom is the **sternum**, and the lateral, or side sclerites, are the **pleura** (singular **pleuron**). In most insects, the lower corners of each box are reinforced by internal struts; this provides support for the legs and the muscles that power them. In most insects, the thoracic segments are more or less fused into a fairly rigid structure, to support all the leg and wing muscles the thorax houses.

If the insect has legs (some don't), it will have one pair per thoracic segment—hence the six legs that are one of the hallmarks of the insects. The legs articulate with the thorax low on the pleural sclerites. Each leg has five segments; from the base, at the thorax, out, these are the **coxa**, **trochanter**, **femur**, **tibia**, and **tarsus** (Fig 5.2). The tarsus itself has "segments"- usually three to five.

**Figure 5.1.** Schematic diagram of a generalized insect thorax.

Insect legs can be modified for a number of functions. Grasshoppers and crickets have hind, or metathoracic, legs modified for jumping (Fig. 5.3); we call this sort of leg a **saltatorial** leg (this word has the same root as summersault). Legs that are modified for digging, as on this mole cricket (Fig. 5.4), are called **fossorial** legs; this word, meaning, "pertaining to digging," has the same root as the word fossil, which we obtain, oddly enough, by digging. The oar-like, swimming appendages on aquatic insects (Fig. 5.5) are called **natatorial** legs. Mantids (Fig. 5.6) and some other carnivorous insects have **raptorial** legs modified for capturing and retaining prey. Fleet-footed insects like tiger beetles (Fig. 5.7) have **cursorial** or running legs. In many insects, energy is stored in the joints of the legs by bands or balls of rubbery resilin, enhancing the efficiency of the leg muscles.

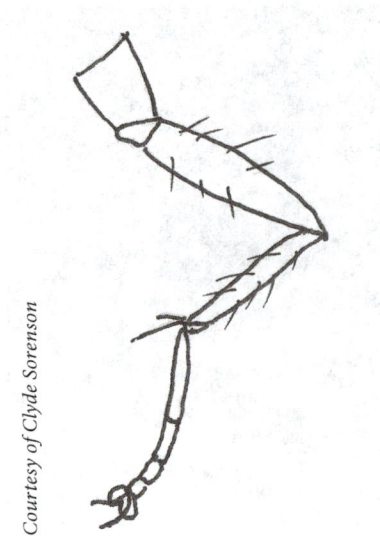

**Figure 5.2.** Schematic diagram of a generalized insect leg.

**Figure 5.3.** The saltatorial hind (metathoracic) legs of a grasshopper.

**Figure 5.4.** The fossorial front (prothoracic) legs of a mole cricket.

**Figure 5.5.** The natorial metathoracic legs of a backswimmer.

**Figure 5.6.** The raptorial prothoracic legs of a mantis.

34  Insects and People

If the insect has wings (many don't), they will be located on either the mesothorax, the metathorax, or both. Wings are never found on the prothoracic segment! The wings articulate at the junction of the segment's notum and pleuron; there's usually a small outgrowth of the pleuron that functions as a pivot for the wing. We'll be discussing this a bit more in a later chapter, but for now, it's important to note that all the muscles that power and direct the wings are located in the thorax. No insect has any muscles in the wing proper. Functional wings are found only in adult insects (with one exception, which we'll discuss later), so if an insect flies by you, you can be sure it won't be growing any more! As with mouthparts, the type of wings an insect has can help us identify the order it belongs to.

The ancestral, primitive form is the **membranous** wing, typified by those found on dragonflies (Fig. 5.8). These wings are usually transparent (like cellophane), although they sometimes are patterned. More primitive wings tend to have large numbers of veins; more advanced insects have wings with reduced venation (Fig. 5.9).

The forewings of many insects like cockroaches and grasshoppers (Fig. 5.10) are thickened and leathery, and serve to protect the membranous hind wings and abdomen of the insect. This type of wing is called a **tegmina**. In these insects, the large, fan-shaped hind wings do the bulk of the work of flight.

**Figure 5.7.** The cursorial legs of a tiger beetle.

**Figure 5.8.** The membranous wings of a skimmer dragonfly. Note the large number of veins and cross-veins in the wings.

**Figure 5.9.** The membranous wings of a bumblebee. Note the reduced venation compared to the dragonfly.

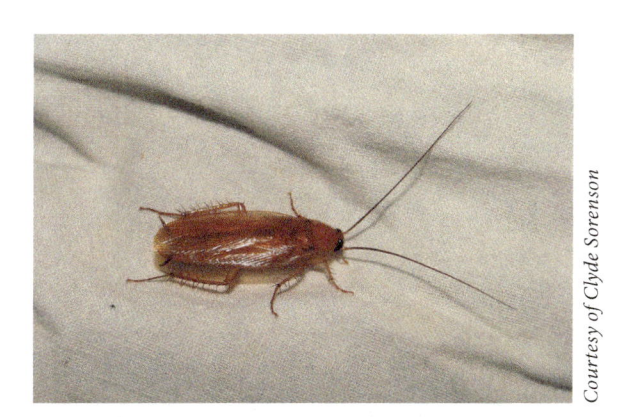

**Figure 5.10.** The tegmina (thickened, leathery forewings) of a wood roach.

The Rest of the Story: The External Anatomy of the Thorax and Abdomen

In the thrips, those weird little insects that have the asymmetrical rasping-sucking mouthparts, the wing (they have four) is essentially a rod of chitin with long setae on two edges; they bear a remarkable resemblance to a small bird feather (Fig. 5.11). These odd wings give the order the thrips belong to its name; *Thysanoptera* means "fringe winged."

True bugs (part of the order Hemiptera), like the stink bug (Fig. 5.12), have forewings which are leathery and thickened on the part closest to the body, and membranous on the distal (away from the body) end. These wings are called **hemelytra**; when folded, they produce a characteristic "X" pattern on the insects that makes them easy to recognize. The leathery inner part again serves to protect the hind wings and the abdomen as the insect navigates through vegetation.

**Figure 5.11.** Thrips showing fringed wings.

The forewings of beetles (order Coleoptera) have been given over almost entirely to a protective function; in most, they have become greatly thickened and heavily sclerotized, and have lost all traces of venation (Fig. 5.13). These **elytra** form a hard, protective "shell" over the large, membranous hind wings and the abdomen. The extraordinary protection these elytra provide is one of the reasons beetles are so incredibly diverse and abundant; wings such as these allow beetles to live in virtually any habitat and exploit virtually any resource while still maintaining the ability to fly. They can tunnel through soil, wood, or even dung; they can live in sand dunes or under water, and they can pursue prey through vegetation, galleries in wood, or coarse gravel, without damaging their delicate hind wings.

Butterflies and moths (order Lepidoptera) have membranous wings clothed in tiny, often colorful chitinous scales (Fig. 5.14). *Lepidoptera* means "scale-wings." While these scales serve many diverse functions in modern butterflies and moths, they probably originated in primitive moths as a means of escaping spider webs. The scales stick to the webbing and then pull off, allowing the moth to escape. They still serve this function in many modern moths. The closely related caddisflies of the order Trichoptera (Fig. 5.15) have wings clothed in "hairs" (setae). *Trichoptera* means "hair-wings."

**Figure 5.12.** The hemelytra of a stink bug.

**Figure 5.13.** The elytra of a long-horned flower beetle.

36  Insects and People

**Figure 5.14.** The scale-covered wings of a cecropia moth.

In many flying insects, the forewings and the hindwings are linked together, turning two wings on each side into essentially, one wing. In Hymenoptera, this is accomplished by a row of tiny hooks on the leading edge of the hind wing (Fig. 5.16) that engage the trailing edge of the forewing. We'll discuss the reasons why in a later chapter.

**Figure 5.15.** The setae-covered wings of a caddis fly.

# The Abdomen

The head is dedicated to sensing the world and ingesting food, and the thorax is dedicated to locomotion. The **abdomen** (Fig. 5.17) is the seat of most of the other functions necessary to sustain insect life and reproduction. The abdomen contains the bulk of the digestive system, the excretory system, much of the circulatory system, and the reproductive system. The abdomen contains between six and eleven segments, with more, typically, in more primitive insects, and fewer in more derived insects. Given that the head is a consolidated capsule, and the thorax is a fairly rigid, three segment box, the abdomen in most insects is often fairly flexible, giving the animal enhanced maneuverability. In most insects, there are relatively few appendages found on the abdomen, but it is important to remember that the ancestral form that gave rise to the insects probably had one pair of walking appendages on each of the segments that would eventually become the insect abdomen.

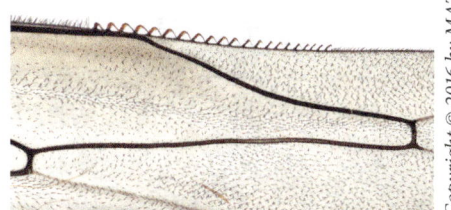

**Figure 5.16.**

**Figure 5.17.** The abdomen of a dragonfly.

The most prominent appendages on the abdomens of most insects are a pair of **cerci** near the back end (Fig. 5.18). The cerci, like the mouthparts, are thought to be derived from ancestral walking appendages, and, in most insects, they are sensory organs that inform the insect of the goings-on behind it. In cockroaches, the cerci are particularly sensitive to air movements (Fig. 5.19). They are connected by massive nerve cells that lead directly to the ganglia in the thorax that control the legs. When the nerve-rich hairs on the cerci detect air movement, the insect immediately starts running away from

The Rest of the Story: The External Anatomy of the Thorax and Abdomen

**Figure 5.18.** A mayfly with long cerci.

**Figure 5.19.** The air current-sensitive cerci of a smoky-brown cockroach.

the source of the moving air, explaining why it is so hard to mash a cockroach with your shoe or a rolled up paper. The roach's brain doesn't get into the game until the critter is already at full speed! In the earwigs of the order Dermaptera (Fig. 5.20), the circi are large and forceps-shaped, and are probably used for defense. The fact that they are larger in the males of some species suggests that they may also be important in sexual selection as well. In most other insects with prominent cerci, the organs probably are primarily chemosensory in function, and in most, the chemicals they sense probably have something to do with the process of copulation.

The most primitive insects tend to have more abdominal appendages. The silverfish and firebrats of the primitively wingless Thysanura often have a peg-like **stylus** (plural **styli**) on either side of each abdominal segment (Fig 5.21). These are thought to derive from the ancestral walking appendages of these segments and appear to have a sensory function. Thysanura and the mayflies of Ephemeroptera also have a long, thread-like **median filament** on the tail between the cerci (Fig. 5.22).

Aquatic insects frequently have external **gills** they use to harvest oxygen from the water they live in. In mayfly nymphs, the gills are feathery outgrowths found on either side of each abdominal segment (Fig. 5.23). Mosquito larvae have a "snorkel" at the tip of their abdomen composed of two extended spiracles (Fig. 5.24). Damselfly nymphs have three large, plate-like gills at the apex of their abdomens (Fig. 5.25). We'll discuss how these things function in the next chapter.

**Figure 5.20.** The forceps-like cerci of an earwig.

**Figure 5.21.** Silverfish. Note styli on sides of abdomen.

38   Insects and People

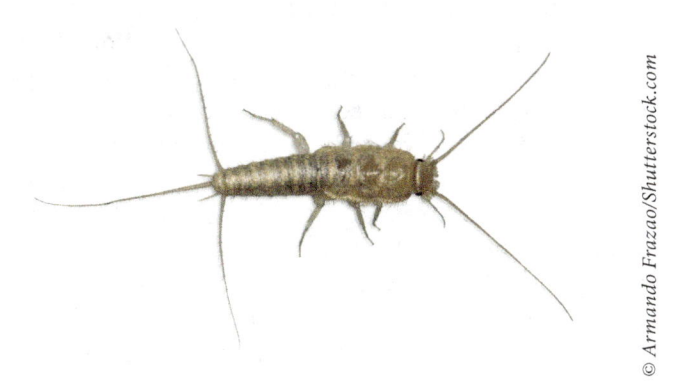

**Figure 5.22.** Median filament, with cerci, of a silverfish.

Most adult female insects have a prominent **ovipositor** which they use to position their eggs as they lay them. The shape of the ovipositor is related to where they lay their eggs: saw-like for sticking them into the wood of twigs (Fig. 5.26), sword-shaped to deposit them deep in the soil (Fig. 5.27), dagger-like to place them

**Figure 5.23.** The underside of an American sand-burrowing mayfly nymph showing the white, hirsute abdominal gills.

**Figure 5.24.** Mosquito larva showing breathing "snorkel."

**Figure 5.25.** Damselfly nymph with plate-like terminal gills.

**Figure 5.26.** Female periodic cicada ovipositing into wood of a twig.

**Figure 5.27.** A female meadow grasshopper with long, sword-shaped ovipositor.

The Rest of the Story: The External Anatomy of the Thorax and Abdomen   39

**Figure 5.28.** Female ichneumon wasp with ovipositor for laying eggs in host insects.

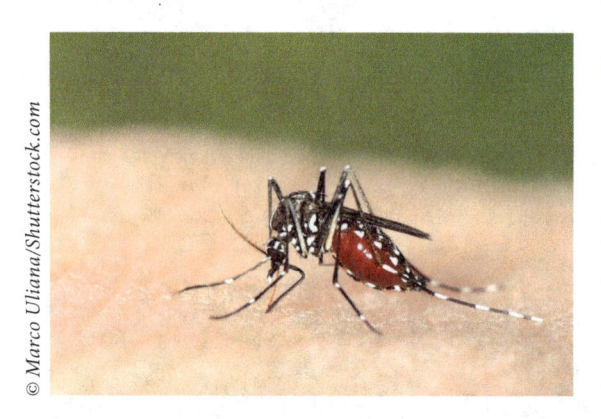

**Figure 5.29.** Engorged female mosquito, showing distended abdomen and inter-segment membranes allowing for abdominal expansion.

**Figure 5.30.** Queen termite.

inside the bodies of other insects (Fig. 5. 28). In many ants, bees, and social wasps, the ovipositors have been converted by evolution into defensive weapons—the infamous stingers these insects wield when threatened. If you get stung by one of these creatures, you know one thing for sure: it was female! Adult male insects have sexual appendages at the tip of the abdomen- the aedeagus (analogous to the penis in the mammals), claspers which the insect uses to ensure it couples long enough to successfully mate, and, often, other structures used to distribute pheromones.

In some insects, the abdomen can be greatly expandable, through the incorporation of extensive membranes between the **sclerites** (plates) of the abdominal segments (Fig. 5.29). This allows these insects to take very large meals quickly when the opportunity arises and then to digest them slowly in a secluded and safe place. Queen termites are exceptional among all insects, in that their abdominal cuticle (exoskeleton) actually continues to grow, without molting, as they assume the role of egg producer for their colony (Fig. 5.30). This is the *only* case in the entire diversity of insects where an insect can grow larger without molting!

40   Insects and People

# VI

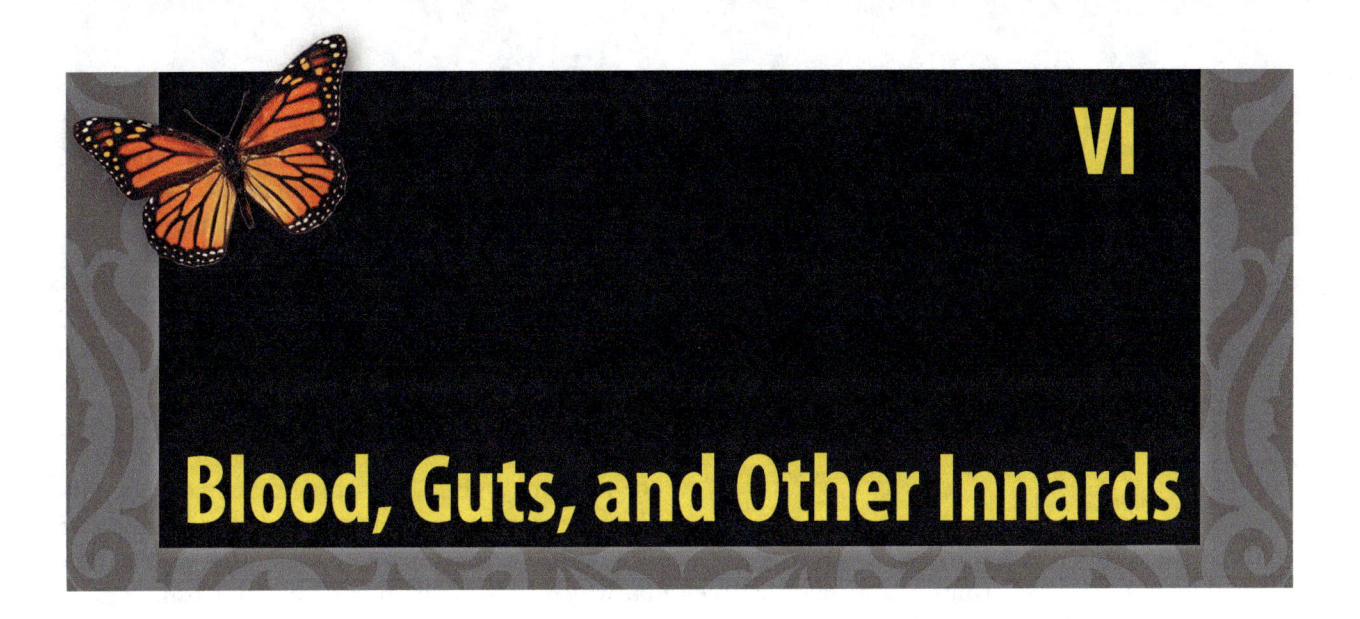

# Blood, Guts, and Other Innards

We've looked at the exterior of the insect, and we now have a good understanding of its exoskeleton and basic body plan. Let's now begin investigating how they go about maintaining life by discussing some of their major organ systems.

Insects are pretty much like virtually all other animals in terms of the basic things they have to accomplish to survive. They must breathe, they must eat, they must deal with toxins in their bodies, and they must prepare for their futures. So it's not too surprising that, in many important details, the ways insects accomplish their basic life functions aren't too different from those of many other animals. But as you should by now expect, there are also many important differences, too. We'll look at the major parts of each of these systems and we'll try to highlight the similarities and the differences.

## The Insect Respiratory System

Insects, like all animals, must take in oxygen and vent off carbon dioxide ($CO_2$) to support the cellular respiration of their tissues. Oxygen is necessary to convert the chemical energy found in nutrients (perhaps a carbohydrate like sugar) into adenosine triphosphate, or ATP, the fuel for cellular function. Carbon dioxide is a waste product of this process; it's important to remember that carbon dioxide is poisonous to animals. Of course, the source for the oxygen insects need, and the "dump" for the waste $CO_2$, is the atmosphere; the modern atmosphere contains about 20% oxygen by volume, with most of the remaining 80% nitrogen. The living cells of an insect's body, as it turns out, have a much more intimate relationship with the surrounding atmosphere than do the vast majority of ours, since insects exploit a **tracheal respiratory system**. In this sort of system, every living cell in the insect's body has a direct connection with the outside world through a series of pipes, the **trachea**, and their finer elements. Most of our bodies' cells "know" nothing of the outside world, since the oxygen and other resources they need are conducted to them by our closed circulatory system. This has a significant bearing on how insects relate to the world.

If we start at the outside and work our way in, the first respiratory structure we encounter is the spiracle (Fig. 6.1). Spiracles are, more or less, holes in the insect's exoskeleton that open to the trachea. In most insects, there is one pair of spiracles on most abdominal segments, and often one pair on the meso- and metathoracic segments, as well (Fig. 6.2a). Spiracles may have a perforated plate

**Figure 6.1.** Luna moth caterpillar. The oblong, brown structures along the animal's side are its spiracles.

to screen dust and to limit water loss; they often have valve-like structures that allow them to be closed and opened, and, in many insects, they are surrounded by long setae, again, to filter dust and reduce water loss. The muscles that open and close them usually ring the spiracle and close it when they contract; the normal, resting, state is open. The ability to open and close the spiracles is critical to the function of the respiratory system, as we shall see in a bit.

The spiracles open to the **trachea**, which are essentially invaginations of the insect's exoskeleton (Fig. 6.2b). Since they are derived from the exoskeleton, the linings of the major trachea are shed along with the rest of the cuticle when the insect molts. The trachea are reinforced with rings of cuticle called **taenidia**, which prevent the tubes from collapsing under negative pressure, much as the coiled wire in a dryer vent hose prevents its collapse. The trachea are interrupted in places by unreinforced **air sacs** (Fig. 6.2c), which can expand or collapse as pressure changes in the insect's body. The trachea divide into smaller and smaller, branching elements, eventually creating a dense network of very fine, very thin-walled elements called **tracheoles** (Fig. 6.2d). The tracheoles are fluid filled when the insect is at rest, and essentially every single living cell in the insect's body lies in close proximity to one of them. Air is forced in and out of the tracheal elements by coordinated contractions of abdominal muscles, in concert, sometimes, with opening and closing of the spiracles. Dorso-ventral muscles attached to the top and bottom abdominal tergites (exoskeleton plates) can contract. When these muscles relax, the natural tension in the exoskeleton causes inspiration; when the muscles contract, the insect exhales. Oxygen is transferred to the living tissues by diffusion from the air in the trachea to the liquid in the tracheoles and then to the cells. When the insect is really active, the fluid leaves the trachea and oxygen diffuses directly from the air in the tracheoles to the cells. $CO_2$ moves by the same processes in the other direction, out of the cells.

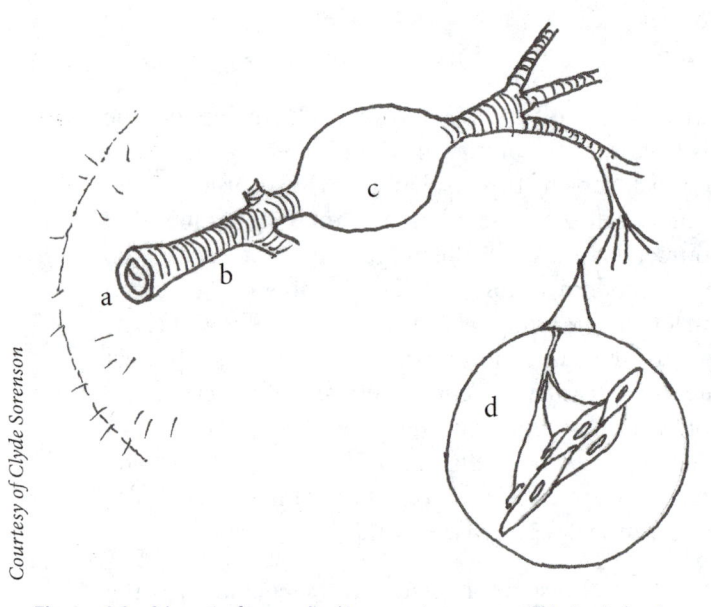

**Figure 6.2.** Schematic of a generalized insect respiratory system. This view is showing only a single spiracle (a) and its related structures: b. trachea; c. air sac; d. tracheoles ventilating individual cells.

In most higher insects, the main tracheal trunks leading from the spiracles are linked by cross trunks so that if a spiracle becomes blocked, oxygen can still be provided to the region of the body normally served by the obstructed opening. The air sacs, in addition to serving as air reservoirs, are important in regulating buoyancy in aquatic insects, and important in creating the internal pressure needed to rupture the old cuticle when the insect molts.

Aquatic insects also breathe through a tracheal respiratory system, but there are several different strategies for delivering oxygen to these air-filled tubes. Some insects, like mosquito larvae, have a snorkel they use to maintain contact with the atmosphere. The snorkel is essentially a long tube with two spiracles at the apex. The spiracles are

42   Insects and People

**Figure 6.3.** Diving beetles with plastron of air.

surrounded by a ring of long, hydrophobic (water-repelling) setae. When the insect is "hanging" at the water's surface, these setae flair out and the insect breathes in a more or less conventional way. When the insect dives, the setae fold over the spiracles, preventing water from entering. This is an excellent way to provide oxygen to an aquatic insect living in the tepid, oxygen-poor waters of stagnant pools. One way we can control mosquito larvae is by spreading a thin layer of oil over the surface of mosquito-laden puddles; the oil destroys the hydrophobicity of the setae, and the insect drowns.

Other aquatic insects, like the diving beetles (Fig. 6.3), actually carry an "aqualung" under water with them. The surface of their abdomen, including the space under their wings, is clothed in short, hydrophobic setae, which trap an air bubble called a **plastron** around the insect's body, including the area surrounding the spiracles. While the air bubble itself does contain some oxygen when it first forms, most of the oxygen the insect actually consumes moves from the water into the air bubble by diffusion as the oxygen in the plastron is depleted; $CO_2$ moves the other way by the same mechanism. The insect breathes normally through its spiracles in this little trapped pocket of atmosphere, and it can remain submerged far longer than would be suggested by the size of the plastron.

Still other aquatic insects have large, thin-walled, external extensions of the trachea commonly called gills. These serve to increase the surface area through which gasses can diffuse into and out of the air-filled trachea. Since cold water contains more dissolved oxygen than warm water, this kind of aquatic respiratory system is very common in insects that live in cold water systems, like the mayflies (Fig. 6.4) and stoneflies (Fig. 6.5) so important to trout and trout anglers, like myself.

**Figure 6.4.** Mayfly.

**Figure 6.5.** Stonefly.

Blood, Guts, and Other Innards 43

# The Insect Circulatory System

While insects don't use their circulatory systems to transport oxygen and carbon dioxide, as we do, they still have need to move many other things efficiently around the interiors of their bodies. Insects have an **open circulatory** system (Fig. 6.6), in which essentially the entire interior space of the body is part of the circulatory system. This interior cavity is called the **hemocoel**, and it is filled with a clear, greenish or yellowish liquid, the **hemolymph**, which is the insect's blood. Hemolymph contains nutrients, salts, defensive cells called **hemocytes**, waste products, and hormones, and it usually is somewhat thicker in consistency than water. The major conductive structure in the circulatory system is the **dorsal vessel**, which, as you might guess from the name, runs along the back, or dorsal side, of the insect. The dorsal vessel is divided into two segments: the perforated **dorsal heart**, which runs through the abdomen, and the aorta, which passes through the thorax and opens as a simple tube into the head capsule. The dorsal heart is actually a series of open-sided chambers that, through rhythmic contractions, force hemolymph forward into and then through the aorta, so that hemolymph circulates forward dorsally and rearward, more or less, ventrally (Fig 6.6). In the legs and wings, there may be **diaphragms** or **accessory hearts** that steer hemolymph through these appendages.

In addition to transporting things, the hemolymph has other important functions. It serves as a hydraulic fluid during wing expansion at molting, it lubricates joints, and, importantly, it can transport heat generated through muscular activity from one part of the body to another.

# The Insect Digestive System

Insects, like most (but not all!) animals, are basically tubes within tubes that process food. Insect guts are compartmentalized into three distinct regions: the **foregut**, **midgut**, and **hindgut** (Fig. 6.7). The foregut and the hindgut, like the trachea, are basically invaginations of the insect's exoskeleton. They are lined with cuticle, and, like the rest of the cuticle, are shed when the insect molts.

Since chewing mouthparts are the ancestral type of mouths, we'll focus first on the digestive system of mandibulate insects. The foregut is further subdivided into several important structures. The **buccal** or mouth **cavity** (Fig. 6.7a) is immediately inside the mouthparts. Usually associated with the buccal cavity are **salivary glands** that produce lubricant to ease the passage of food into the rest of the digestive system; these glands are associated with the labial segment of the mouth. A short esophagus leads to the crop (Fig. 6.7b), where food is stored temporarily before digestion. Most insects, like many other animals, tend to eat as much as they can in a short time and then retreat to a safe place to digest the meal. At the rear of the crop is a structure called the **proventriculus** (Fig. 6.7c); this is, in insects that eat fibrous foods, a circular, heavily sclerotized set of "teeth" that tear the food into tiny particles. The proventriculus is analogous to the gizzard in birds, and it prepares food for actual digestion. Behind the proventriculus, at the junction between the foregut and the midgut, is a muscular cardiac valve that regulates the passage of food between the two regions.

The **midgut** (Fig. 6.7d) is where the actual work of digestion occurs. The midgut is made of epithelial cells that secrete digestive enzymes and chemicals, and absorb the released nutrients. It is lined with a

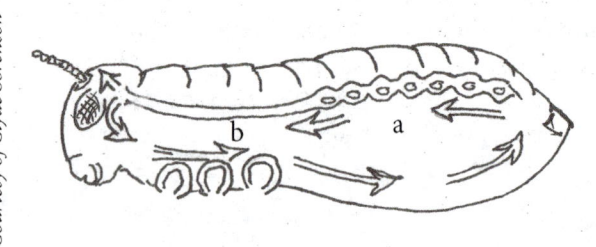

**Figure 6.6.** Schematic of generalized insect circulatory system, showing the dorsal vessel with its heart (a) and aorta (b).

44   Insects and People

polysaccharide (sugar) sheet, the **peritrophic membrane**, which protects the midgut epithelium from sharp food fragments and microorganisms that might be living in the food, and it is shed and passed out with food remains frequently. In many insects, particularly the Lepidoptera, some flies, and some beetles, the midgut contents are very alkaline, or high in ph. This is pretty much the opposite of our gut environment; the human stomach, and that of most other animals, for that matter, is very acidic, or low in ph. Once as much nutrition as possible has been harvested from the food, what is left passes through a second muscular valve, the **pyloric valve**, into the hindgut.

The hindgut (Fig. 6.7e) is, again, lined with cuticle and molted along with the rest of the cuticle. The remaining, undigested bits of the food, together, often, with gut bacteria, are packaged for elimination in the hindgut. Since water is extremely valuable to very small terrestrial animals like insects, much of the water remaining in the waste, together with salts, is resorbed in the hindgut, so that, in most insects, the resulting waste is quite dry. We call the digestive waste product of insects that feed on plants, with chewing mouthparts, **frass**. Frass pellets on the ground are a sure sign of some kind of foliage-feeding insect in overhead vegetation (Fig 6.8). Frass and other waste products are eliminated through the insect's anus.

Insects that feed on liquid substances, like plant sap or blood, often have the opposite problem. They ingest far more water than they need in order to acquire enough nutrients from their food source. Aphids and other plant-feeding, sucking insects produce a sugary waste product called honeydew (Fig. 6.9). They end up taking not only more water than they need, but more sugars as well, as they seek the amino acids and other nutrients that occur in sap at very low relative concentrations. This substance accumulates on surfaces below the insects, such as other leaves, and, since it is full of sugar, supports the growth of fungi, called **sooty mold** (Fig. 6. 10), and other microorganisms that can compromise the health of the plant by preventing sunlight from reaching the photosynthetic tissues of the leaf. The manna from heaven mentioned in Exodus in the Bible is thought to be the crystalized honeydew of scale insects that feed on desert shrubs; in the arid environment of the desert, the honeydew dries too quickly to support the sooty mold fungi. Insects that feed on blood typically dump some of the clear plasma that they ingest through their anus while retaining the protein-rich blood cells (Fig. 6.11).

Many insects, notably termites and aphids, have symbiotic **microorganisms** living in their guts that help them digest relatively indigestible substances and that produce nutrients not found in their food. These relationships are **mutualisms** in that both the insect and the microbes benefit from the relationship. The insect gets essential help in exploiting its food, while the

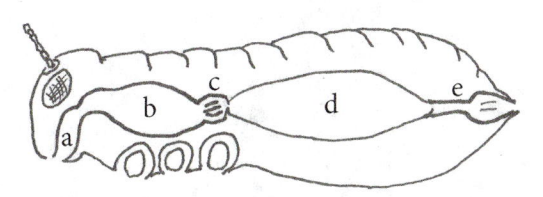

**Figure 6.7.** Schematic of generalized insect digestive system: a- buccal cavity (mouth); b crop; c- proventriculus; d-midgut; e-hind gut with rectum.

**Figure 6.8.** Caterpillar frass.

**Figure 6.9.** Aphids on a leaf. The shiny substance on the leaf surface is honeydew.

Blood, Guts, and Other Innards 45

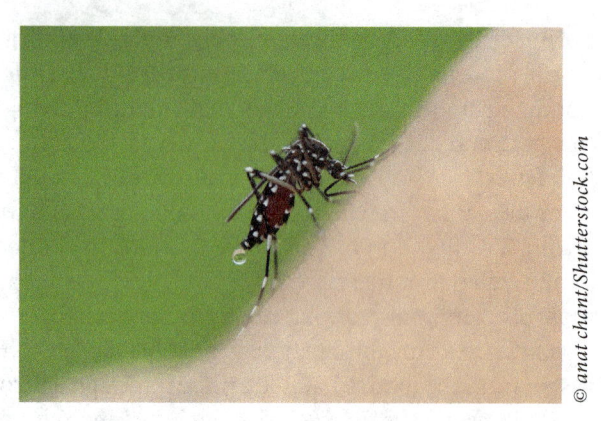

**Figure 6.11.** Blood-feeding mosquito excreting excess plasma.

microbes get a safe place to live and a reliable food source for themselves. The symbionts that live in the guts of most termites are protozoans that have within them symbiotic bacteria that actually digest cellulose. The symbionts in many other insects may be bacteria or fungi. Symbiotic organisms can help insects cope with other kinds of stresses, too, and we may touch on this later in the text.

**Figure 6.10.** Sooty mold on aphid-infested leaf.

# The Insect Excretory System

Insects, like other animals, use carbohydrates as the fuel to run their cells. Many substances, like sugars, are simple carbohydrates, but insects may exploit other substances, like proteins, as well. Proteins are composed of chemicals called amino acids (Fig. 6.12), and amino acids are essentially carbohydrates with nitrogen-containing amine groups attached. Amino acids are essential for the construction of tissues, but they can also be metabolized as energy sources once the amine group is removed. The amine group is very similar chemically to ammonia, and can convert to ammonia if left alone; ammonia is quite toxic to living tissues. The amine fraction thus becomes a nitrogenous (nitrogen-containing) waste product that must be eliminated. In most animals, the amino fraction is converted to a chemical called **urea**, which is fairly water soluble and significantly less toxic than ammonia. We void our nitrogenous waste as urea as a fairly dilute liquid urine. However, remember that, for most terrestrial insects, water is a very rare and precious commodity, and they don't have a lot of it to "waste" as urine. In most insects and several other groups of animals, urea is further converted to another chemical, **uric acid**, which is very low in water solubility, very low in toxicity, and which can be eliminated as a solid, saving a tremendous amount of water. Converting urea to uric acid is metabolically expensive, but the water savings are significant. The organs that remove metabolic waste from the insect's body, analogous to our kidneys, are the **Malphigian tubules** (Fig. 6.13).

The Malphigian tubules in most insects are yellowish or whitish, blind tubes attached to the hindgut just behind the pyloric valve. They extend out into the hemolymph, and the wastes move into the tubules through diffusion and active transport. As the waste liquid moves down the tubule, it is concentrated, so that when it dumps out into the hindgut, uric acid precipitates out and the water can be reabsorbed. In some insects, the ends of the tubules are embedded in the fat body, and the fat body may contribute to excretion.

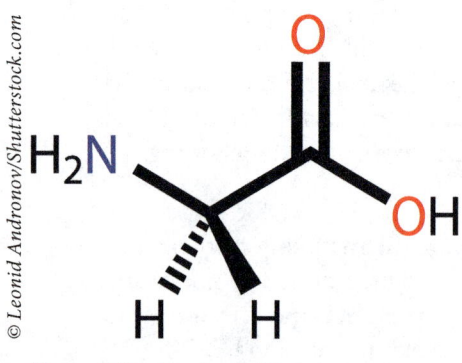

**Figure 6.12.** An amino acid (glycine).

46   Insects and People

Many aquatic insects excrete their nitrogenous waste as ammonia in copious amounts of dilute urine. Aquatic insects essentially have to urinate abundantly, because their relatively "salty" bodies, immersed in freshwater, are constantly absorbing water through osmosis. A few insects excrete urea.

# The Insect Fat Body

Like most animals, insects, through most of their life, need to take in more energy than they require for simple maintenance of life; if the insect is to grow and to reproduce, it needs extra, stored energy for these tasks. Insects sock away the extra energy they harvest from their food in an organ called the **fat body** (Fig. 6.13). Extra sugars can be converted to fats, which are much more energy dense, for storage. In most insects, the fat body is a white amorphous mass that can, in a well-fed insect, occupy the bulk of the hemocoel of the insect. The fat body is also important in hormonal functions, but we'll discuss that in a later unit.

**Figure 6.13.** Dissected cockroach. Fine white filaments are Malphigian tubules; amorphous white masses are fat body. Silvery filaments are trachea. The gut can be seen towards the top of the specimen.

*Courtesy of Clyde Sorenson*

# The Insect Nervous System

Insects, as it turns out, have nervous systems that resemble ours to a very great degree, but there are some important differences. Their nerve cells "work" in exactly the same way ours do; insect nerve cells communicate with each other with some of the same neurotransmitting chemicals. The insect nervous system (Fig. 6.14) is composed of a central nerve cord, peripheral nerves that communicate with the muscles and sensory organs, and the brain. Unlike the central nervous system of vertebrates, the central nervous system of insects is ventral (located along the "belly" of the creature) to the digestive system; ours is dorsal (along the back) with respect to the digestive system. There is a pair of **ganglia** (singular, ganglion), or mass of interconnected nerve cell bodies, in each body segment, and the ganglia in each segment are more or less in control of that segment, while communicating with other segments. This is why a headless insect can continue to walk and run for days, or longer, and why a male preying mantis can continue to successfully mate even if the female has eaten his head off. The activity of the legs is largely regulated and coordinated by the ganglia in the thoracic segments, while the sex organs are directed by ganglia in the abdomen.

The insect brain is, compared to a vertebrate brain, a very rudimentary structure, but it obviously is a successful plan. The brain consists of three pairs of clustered, fused ganglia; the front-most pair is largely responsible for processing vision while the middle pair processes information from the antennae. The third pair communicates with the mouthparts and the rest of the body. Unlike vertebrate nervous systems, there are no ganglia in the appendages, and the nerves are not insulated by a fatty myelin sheath.

There are additional insect life systems we haven't covered in this chapter, but we will get to them in future installments.

Blood, Guts, and Other Innards   47

# NERVOUS SYSTEM OF WALKING STICK

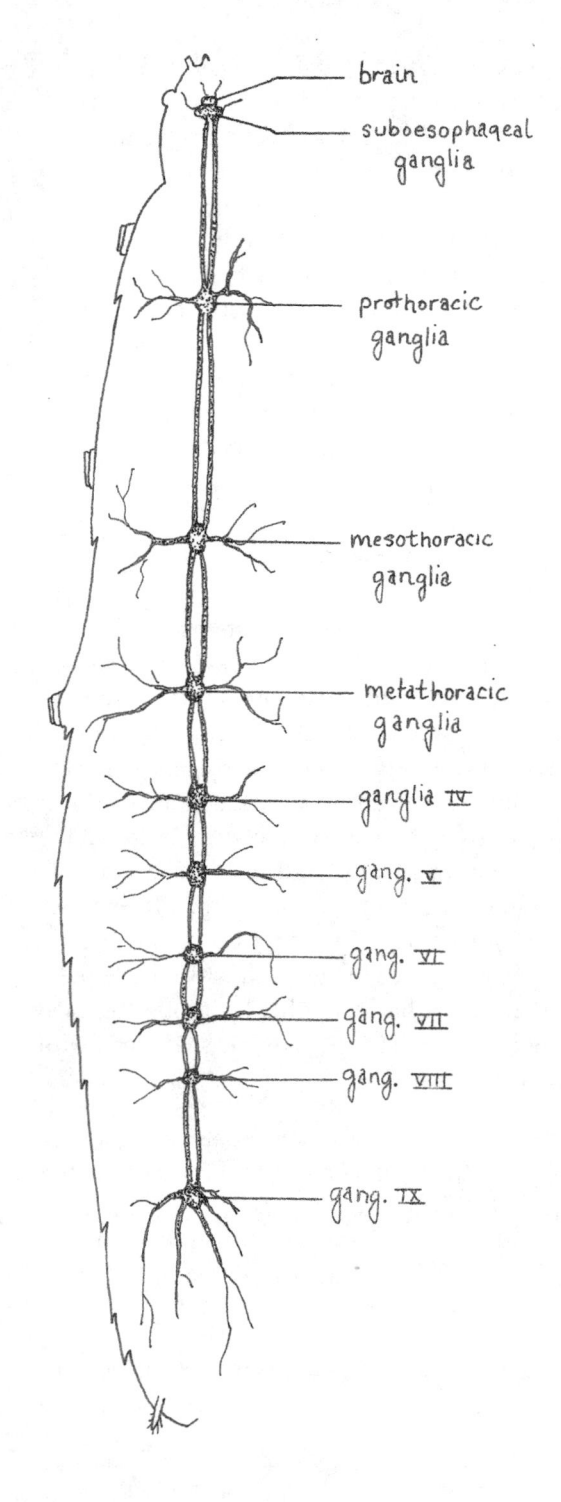

**Figure 6.14.** The nervous system of a walking stick insect.

# VII

# Have You Ever Metamorphosis?
## Insect Growth and Development

An animal that perishes before it reproduces is an evolutionary loser. All animals face tremendous challenges in reaching maturity; a creature must eat and grow, and last long enough while doing this to reach an age at which it can mate and leave offspring to the next generation. Different animal species exploit different patterns of growth and development to ensure their continued survival and, with about a million known species of insects and millions left to discover, it should not be surprising that there is tremendous variability in how insects mature.

The basic life cycle of most insects is much like that of most animals (Fig. 7.1). Like most other animals, most insects start life as a fertilized egg, a product of the combination of an egg cell from the mother and a sperm cell from the father (more about this in a later chapter). Most insects deposit eggs shortly after they are fertilized, and the eggs are usually laid either in, on, or very near whatever comes out of the egg will need as food. After a variable period, the egg hatches, and

**Figure 7.1.** Generalized Animal lifecycle: a- egg; b-immature; c-adult.

an immature insect emerges and generally begins feeding as soon as it can. The main job of the immature, apart from not getting eaten by something else, is to sock away as much nutrition as possible for the future. It undergoes several immature stages (called **instars** in insects) with each new stage defined by a molt of the cuticular part of the exoskeleton (remember that it can't stretch and must periodically be replaced if the insect is to grow). Eventually, once it has fed and grown enough, it molts one last time to the **ultimate instar**, the adult stage. In most insects, the adult is distinguished from the immatures by two important abilities: the ability to fly (only adult insects have functional wings), and the ability to mate and reproduce. If you see an insect fly by, you know that it is an adult. With a couple notable exceptions that we'll get to in a bit, adult insects don't molt and don't get any bigger.

Molting is a critical event in the life of every insect (every arthropod, for that matter), so it's important that we have a solid understanding of how the process works. Remember, the exoskeleton is composed of a thin living layer, the epidermis, and a non-living, chitin-rich cuticle produced by it. Also remember that the cuticle does not stretch, so in order to grow, the insect has to get out of the old cuticle and into a new one. This process starts when stretch receptors in the epidermis detect the strain produced by a too-full body cavity. This sends a message to the brain, which produces a hormone that signals the production of another hormone, **ecdysone**. Ecdysone signals to the epidermis that it is time to produce a new cuticle. To begin the actual process, the old cuticle separates from the epidermis, an event called **apolysis**. The epidermis secretes a new epicuticle, and then secretes a substance called molting fluid in this new void; the molting fluid then starts to digest the old endocuticle. The resulting glucosamine and protein are partially recycled as the epidermis builds first the new exocuticle and then the new endocuticle.

**Figure 7.2.** A cicada emerging from its last nymphal exuvium. Not the thin, transparent character of the old exoskeleton.

Of course, the new cuticle must be larger than the old one if the insect is to grow, so it is usually wrinkled and wadded up under the now much thinner old cuticle (Fig. 7.2). Once the new cuticle is ready, the insect captures as much air as possible in its tracheal system creating internal pressure that causes the old cuticle to rupture along weak seams, usually along the dorsal surface of the thorax, and the creature wriggles out of its old skin. What emerges is soft and usually much lighter in color (Fig. 7.3), because the new exocuticle has not tanned, or **sclerotized**, yet. Phenolic compounds are injected into the exocuticle by the epidermis, and these chemicals combine with and change the proteins in this part of the cuticle so that it stiffens and darkens, producing the sclerotin characteristic of the exocuticle. This process can take just a few minutes to many hours, and since the insect has limited mobility and is particularly vulnerable while it molts, many insect retreat to hiding places and molt under the cover of darkness. Remember, when an insect molts, it sheds its entire cuticle including those parts lining the tracheal respiratory system and the fore- and hindgut.

Insects follow three basic patterns of development, producing three major lifecycle patterns. These are **ametaboly**, **hemimetaboly**, and **holometaboly**.

# Ametabolous Life Cycles: Life With Little Change

The most ancient living insects, those that are most like the very first insects, are regarded as **primitively wingless**; this means that they are always wingless, and all their ancestors were wingless. Perhaps the best example of such a creature is the silverfish (Fig. 7.4), of the order Thysanura. These are the insects that undergo **ametabolous** ("no change") development, which

**Figure 7.3.** Freshly eclosed periodic cicada, show light-colored, untanned cuticle.

50  Insects and People

essentially means that the immatures look exactly like the adults, except for size. In these insects, there are three life stages: the egg, the **young**, and the adult, and, apart from size, the only thing distinguishing the young from the adults is that the adults have functional reproductive organs. Since ametaboly is restricted to a handful of very small orders, the fewest number of insect species undergo this sort of development.

Since they don't have wings, the adults of some of these insects do something that no winged insect does (with one exception): they continue to molt. This trait is most well developed in the silverfish—some may molt 50 or 60 times during their five-year adult lifespan. It is thought that these insects continue to molt as a defense against disease organisms that colonize their cuticle; molting also allows them to replenish the cuticular scales that clothe their bodies, protect them from spider webs, and give them their silvery appearance.

It is important to remember that, since the young and the adults are essentially identical in form, they have the same kinds of mouthparts and live in the same kinds of habitats, so they essentially compete with each other.

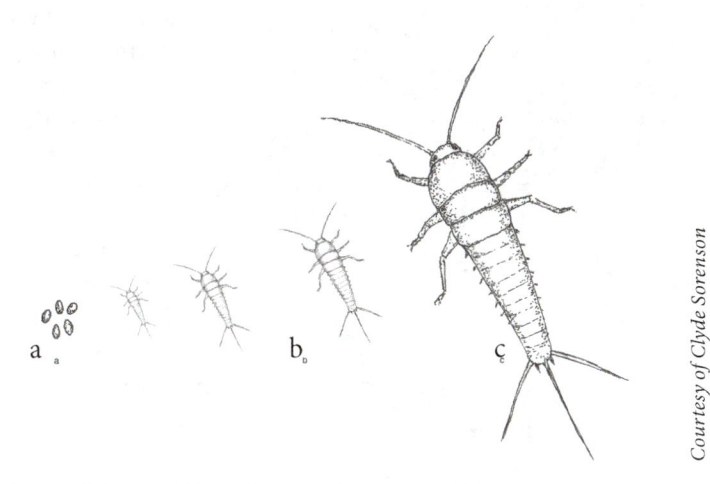

**Figure 7.4.** Silverfish lifecycle- an example of an ametabolous insect. a- eggs; b- young; c- adult.

# Hemimetabolous Life Cycles: Wings Growing on the Outside

The next evolutionary step up from ametaboly is **hemimetaboly**, which is often called partial, simple, or incomplete metamorphosis. This is the form of development found in many of the more ancient winged insects, like dragonflies and mayflies, but is also found in the Orthopteroid orders (grasshoppers, roaches, mantids, and the like), the true bugs of the order Hemiptera, and several other orders. In these insects, there are again three life stages: the egg, the **nymph**, and the adult (Fig 7.5).

In those hemimetabolous insects with terrestrial nymphs, the immatures and the adults usually look very much alike, with similar mouthparts, similar compound eyes, and, usually, inhabiting similar habitats; again, the immatures and adults of hemimetabolous insects often compete with each other for resources. What distinguishes nymphs from adults, apart from size, is the presence of functional wings in adults, and non-functional, but apparent, **wing pads** in nymphs (Fig. 7.6).

The nymphs of those species with aquatic immatures, like the dragonflies, mayflies,

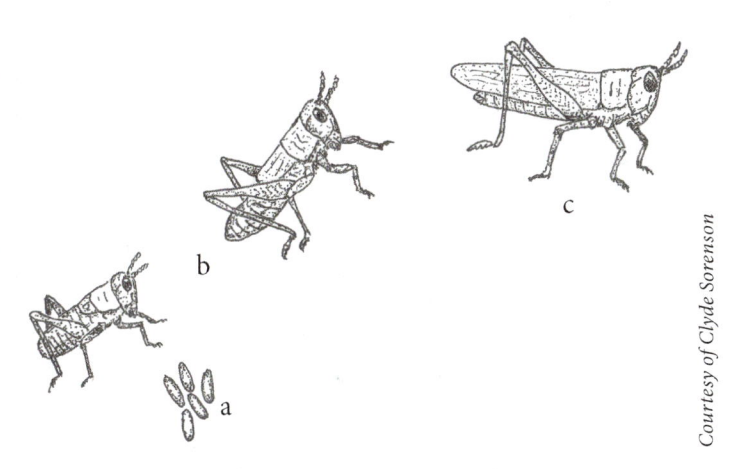

**Figure 7.5.** Grasshopper lifecycle—an example of a hemimetabolous insect. a- eggs; b- nymphs; c- adult.

Have You Ever Metamorphosis? Insect Growth and Development 51

and stoneflies, often look somewhat different from their terrestrial adult forms, but they still have those wing pads (Fig. 7.7). These aquatic nymphs are often called **naiads** to differentiate them from insect species with terrestrial nymphs. The term *naiad*, by the way, comes from Greek mythology; the Greek Naiads were the feminine spirits that guarded sacred streams.

The wing pads of hemimetabolous nymphs get proportionately larger with each successive molt so that one can ascertain the nymphal instar in many species by the size of the wing pads. Once these insects reach full adulthood, they do not molt ever again. If you see an insect fly by, you know it is an adult and will not get any larger.

While a large number of ecologically and economically important insects undergo incomplete metamorphosis, only about 15% or so of all insect species exhibit this development pattern.

**Figure 7.6.** Stink bug nymph. Note wing pads.

**Figure 7.7.** Dragonfly nymph. Note the wing pads on the back of the thorax.

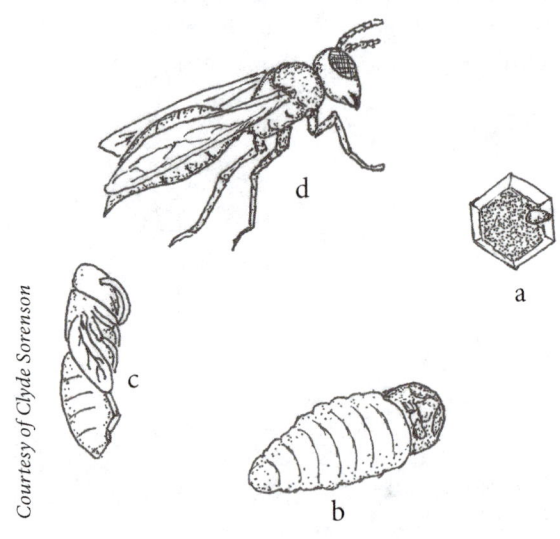

**Figure 7.8.** Paper wasp lifecycle- an example of a holometabolous insect. a- egg in paper nest cell; b- large larva; c- pupa; d- adult.

# Holometabolous Life Cycles: Different Stages, Different Lives

The vast majority of insect species, about 85% or so, undergo **holometaboly**, or complete metamorphosis, for some very important reasons we'll get to in a bit. In this kind of life cycle, there are four life stages (Fig. 7.8): the egg, the **larva**, the **pupa**, and the adult. The huge evolutionary innovation in holometaboly is the pupa. This stage essentially allows everything that was once larva to reassemble into a radically different adult. Consequently, the larvae and adults of a great many holometabolous insects have completely different mouthparts and live in completely different habitats, thus allowing them to exploit patches of habitat far too small to support both adults and immatures. This is, perhaps, the greatest advantage of complete metamorphosis: the ability to lead radically different lives as immatures and adults. The four "mega orders," the Coleoptera (beetles), Lepidoptera (moths and butterflies), Diptera (true flies), and Hymenoptera (wasps, bees, and ants), each containing well over 100,000 species, are all holometabolous, as are several other orders. All holometabolous insects, by the way, are descended from a common ancestor.

In the vast majority of holmetabolous species, the larvae look nothing like the adults, and, indeed, based purely on appearance, and without any foreknowledge (like that you are getting here), it might

52   Insects and People

be impossible to predict what most larval insects will be when they grow up. A caterpillar bears no resemblance to a butterfly (Fig. 7.9), and a maggot looks nothing like a housefly (Fig. 7.10). These remarkable transformations are due to a fundamental difference in the way holometabolous insects develop, compared to the others. In hemimetabolous insects, the structures that will become the final, adult form develop gradually with each successive nymphal molt, but in holometabolous insects, the larva retains its larval form for several instars before rather dramatically assuming a pupal form, and then even more dramatically, the adult form. In these insects, most of the structures that will define the adult form are present in the larvae only as patches of cells called **imaginal discs**, which remain dormant until the pupal stage. In the pupal stage, virtually all the tissue that was the last stage larva dissolves into a "soup" of nutrients that the imaginal discs use to grow into the adult structures. The larva's sole mission, apart from not getting eaten before it finishes development, is to sock away as many nutrients as possible in as brief a period as the environment will allow to fuel this transition.

Larval insects, with virtually no exception, have no compound eyes; any vision they have is restricted to the ocelli-like stemmata, and many larvae have no visual sense at all. In most holometabolous larvae, the antennae are extremely reduced, because they don't have much need for long distance chemoreception. They often have fairly thin and pliable cuticles that are economical to make and easy to molt, so their form is often quite simple-looking. That being said, larvae do come in a remarkable number of forms (Fig. 7.11a-d). They never have anything that looks even remotely like external wing pads.

**Figure 7.9.** Monarch butterfly caterpillar (a.) and adult (b.).

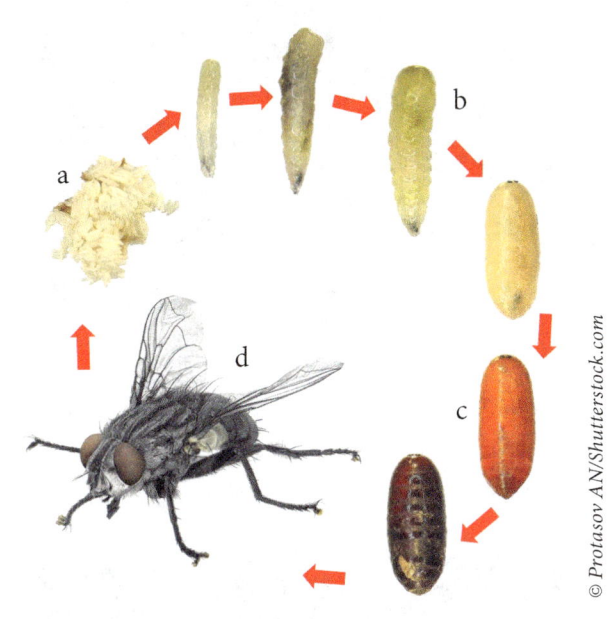

**Figure 7.10.** Housefly eggs (a), larva (maggot) (b), puparium (c), and adult (d).

The pupae of insects that undergo complete metamorphosis are typically incapable of very much movement (apart from sometimes vigorous wriggling), don't feed, don't eliminate waste, don't have much in the way of sensory capacity, and don't appear to do very much of anything. But, of course, there are big things happening inside that inert case. It is often possible to see shadows of what is soon to emerge from the pupa (Fig. 7.12a–e). You can often see where the wings, legs, antennae, and other structures are developing within the pupal cuticle. You can often even determine the sex of insects at this stage. In many holometabolous insects, notably some of the moths and Hymenoptera, the larvae spin a protective silken cocoon just before they molt from the last larval instar into their pupal cuticles.

**Figure 7.11.**  a. Beetle larvae; b. fly larva; c. ant larvae; d. swallowtail butterfly larva.

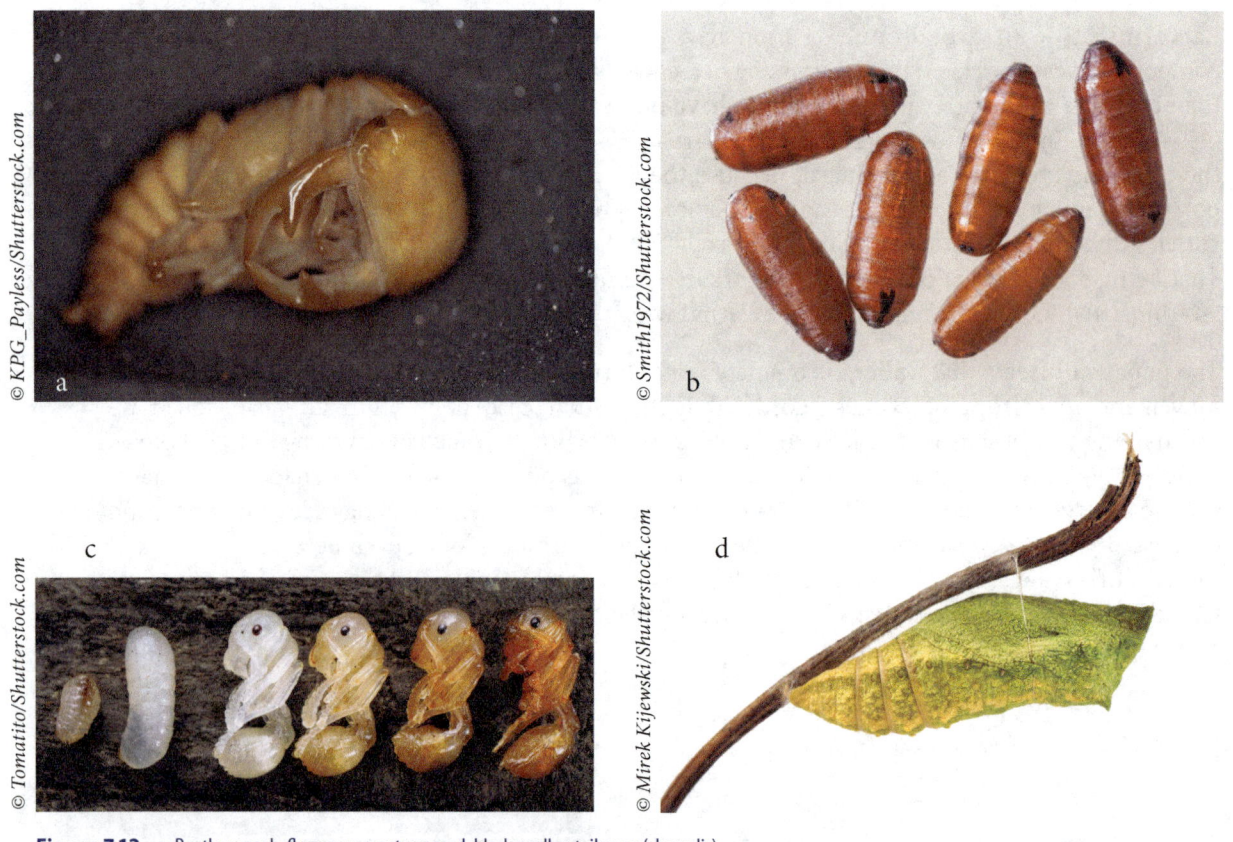

**Figure 7.12.**  a. Beetle pupa; b. fly pupae; c. ant pupae; d. black swallowtail pupa (chrysalis).

54   Insects and People

The extreme compartmentalization of life functions in many holometabolous insects also extends to the adult stage. The primary functions of the adult are to disperse and to reproduce, and many holometabolous adults are basically flying reproductive organs. They don't feed and are driven solely by the urge to reproduce.

# The Long and the Short of Insect Life Cycles and Insect Generations

We generally think of insects as quite short-lived animals, but, as we've come to expect, such generalizations are rash with such a diverse group of animals. A life cycle is the amount of time that must pass from one life stage to the same life stage in the next generation. With insects, it's often convenient to think in terms of the span of time from eggs in one generation to eggs in the next. Insect life cycles range in duration from as little as five days in some aphids (Fig. 7.13), which reproduce parthenogenically and give birth to live young (more on that later), to at least 17 years in three species of periodical cicadas (Fig. 7.14). Some individuals of certain wood boring beetles have documented lifespans of at least 51 years, although this is not the norm for these species. So some insects can live for very long times indeed. However, most insects that we are familiar with have life cycles measured in weeks or months. Just how long an individual insect, in an individual species', lifespan is, however, not that straightforward, since insect growth and development is, at its essence, a complicated network of chemical reactions, and the speed of chemical reactions is dictated by the temperature those reactions occur in.

The internal temperature of most insects is regulated by the temperature of their surroundings. They are therefore **poikilothermic**, meaning their temperature fluctuates with that of their surroundings (however, there are a great many exceptions we'll get to later). Many insects are also **ectotherms**, deriving most of their heat from their surroundings. Humans, by comparison, and like almost all other mammals and all birds, are **endothermic homeotherms**, meaning we maintain a stable internal body temperature by metabolically generating heat. Since the body temperature of insects fluctuates with their environment, the rates of their growth and development also fluctuate; insects of a particular species will grow much slower under cool temperatures than under considerably warmer temperatures. The same aphid that turns a generation in five days during the heat of summer might require ten days or a month, to turn a generation under the cool conditions of late March or April.

**Figure 7.13.** An aphid giving birth.

**Figure 7.14.** A periodic cicada.

Have You Ever Metamorphosis? Insect Growth and Development

Compounding this variability is the fact that there are limits to the temperatures insects can withstand without perishing. On the lower end of the temperature spectrum, most insects pretty much cease doing anything significant when the temperature reaches about 40° F. Most perish if exposed to temperatures of 0° F for more than a couple hours. On the other end of the temperature spectrum, most insects die if exposed to temperatures in excess of 140° F or so for more than an hour, or temperatures in the range of 125° for more than a couple of hours. However, there are several species of desert-dwelling, scavenging ants that regularly forage when the air temperature in their habitat exceeds 120° F.

We can exploit these temperature limits to help manage pest insects. We routinely refrigerate food to reduce or stop the activity of insects in that food, and we use cold storage to protect other goods like furs. We can eliminate infestations in small lots of grain products or other goods by freezing them for at least 24 hours. We can also use high temperatures to control some pests. One of the best ways to deal with bed bug infestations is to use a heat gun to heat-treat the harborages these pernicious insects utilize.

I just said, above, that most insects can't tolerate temperatures below 0° F for more than a few hours, yet insects are abundant throughout the temperate and even arctic zones on earth. So how is it insects can survive in these areas when winter temperatures often get too cold, ostensibly, for insects to survive? Insects have a great range of ways to cope with inhospitable environmental conditions, and we will get around to discussing most, but one of the most important is a condition called **diapause**.

Diapause is a physiological condition in which development, and, often, activity ceases in response to environmental cues that signal those impending inhospitable environmental conditions I mentioned above. In northern temperate areas, like most of the United States, the most significant recurring inhospitable environment is winter, with its severe cold. The cue that triggers diapause in this part of the world is usually the shortening day length as autumn approaches. Insects entering diapause change; they often accumulate chemicals, like glycerol, in their hemolymph that reduce its freezing point, and they build their fat reserves. The sex organs of adults that enter diapause may atrophy; the epicuticle may thicken to increase protection against water loss. Their behavior also changes; insects entering diapause typically cease feeding and clear their guts of food and waste, and they seek out protected environments where temperature extremes may be moderated. They remain in diapause, with much reduced metabolism, until other predictable environmental cues (like, perhaps, the lengthening days of spring) signal the time to regain activity. Only one life stage diapauses in most insect species- some diapause as embryos in eggs, some as larvae, some as pupae, and some as adults.

Insects in tropical areas may also diapause in response to either dry or wet seasons, extreme heat, or in response to a seasonal loss of food resources. Since day length doesn't fluctuate much in tropical areas, other cues trigger diapause, including, prominently, food quality.

# What Controls it All?

The growth and development of insects is, as in all animals, regulated by the hormones produced by the endocrine system. Insects have exquisite hormonal systems quite different from those of vertebrates. While a lot more goes on in insects than we really need to address here, it is important to understand the basics of hormonal control of metamorphosis in insects, because we can actually exploit this information to help control pest species.

Metamorphosis is basically regulated by the relative balance, or titer, of two hormones. One we've already briefly discussed—ecdysone, or molting hormone. The other is a group of chemicals collectively called **juvenile hormones**. These hormones are produced by a structure associated with the insect brain called the corpus allata. Juvenile hormone, as you might expect from the name, keeps a molting insect in a juvenile form. Let's examine how this system works in an insect that goes through complete metamorphosis (Fig. 7.15). If the epidermis receives a signal that is composed of lots of ecdysone plus lots of juvenile hormone, proportaionally, the new cuticle it constructs will be another larval cuticle. If the epidermis receives a signal that is composed of lots of ecdysone and only a little juvenile hormone, the new cuticle will be a pupal cuticle. If the signal is only ecdysone, the new cuticle will be an adult cuticle. In an insect that undergoes incomplete metamorphosis, the titer of juvenile hormone decreases with each successive molt, until, at the last molt, it declines to the point that the signal is all ecdysone.

Lots of molting hormone + Lots of juvenile hormone =

Lots of Molting hormone + a little juvenile hormone =

Lots of molting hormone + almost no juvenile hormone =

**Figure 7.15.** Hormonal regulation of development in a holometabolous insect.

Juvenile hormones often reappear in adult insects and have a number of very important functions in them. This hormone initiates the development of eggs in females, plays a role in regulating diapause, and in mating and migration behaviors.

The discovery of juvenile hormone is, by the way, a really interesting story of propinquity in science. In the early 1960s, a research group was trying to rear a colony of a true bug, but they could not maintain the colony because it never produced adults; the last stage nymphs just continued to molt into larger nymphs until they died. By a process of elimination, the researchers determined that the problem was the paper toweling they were using as bedding for the insects. The paper towels were manufactured from balsam fir pulp, and, upon further investigation, it was determined that the fir pulp contained a chemical they called "the paper factor" that was keeping the insects immature. This chemical was later isolated and its structure was described; it was called **juvabione** for its effect on insects. Juvabione is produced by the trees to protect them from some of the insects that attack them by interfering with the insects' ability to reproduce. Juvenile hormones were isolated from insects and their structures described about the same time.

We make synthetic versions of both ecdysone and juvenile hormone to use as insecticides. These hormone mimicking insecticides are very safe for vertebrates, including us, since vertebrates don't have these hormone systems. Juvenile hormone mimics like methoprene are very useful for managing some insects, like mosquitoes, that are pests as adults but not as larvae. Applying methoprene to a mosquito-breeding container prevents the larvae from ever metamorphosing to adults. Of course, this strategy would be disastrous for an insect that was a pest in the larval stage, like a crop-feeding caterpillar, since you would grow super-caterpillars that would eat even more! Synthetic versions of ecdysone, such as a chemical called tebufenozide, can be useful for these pests; applying tebufenozide to caterpillars causes them to molt prematurely and with fatal consequences.

# VIII

# Yes, Bees Do It: The Mating Game

Insects are, for the most part, very small animals living in a big, hostile world. Many, many things not only like to eat insects, they depend on them for their very existence. As I said at the top of the last chapter, an animal that doesn't leave offspring to the next generation is an evolutionary loser, and the entire genetic legacy of an animal that fails to reproduce is extinguished with its death. Insects face tremendous pressures to survive, and they harness tremendous reproductive horsepower to withstand the many assaults they confront.

Insects, like most animals, start life as an egg. The basic gear they use to produce and fertilize these eggs is much like other animals, with, as you might expect, some important differences. In order to understand the basic process of insect reproduction, and all the wondrous and fascinating variations on this basic process, we first need to get familiar with their reproductive equipment.

## Pieces and Parts I: The Male Insect Reproductive Tract

As in other animals, the primary organs of reproduction in male insects are the **testes** (singular **testis**) (Fig. 8.1). The testes are the manufacturing plant for the male's reproductive cells, the **haploid** sperm, which are produced through a process called **spermatogenesis**. Haploid cells contain half the number of chromosomes normally found in the cells of a given species. If an organism normally has 23 pairs of chromosomes (total 56) in the cells of its body, then the sperm will contain 23 chromosomes, one from each pair in the normal complement. In the testes, germ cells called

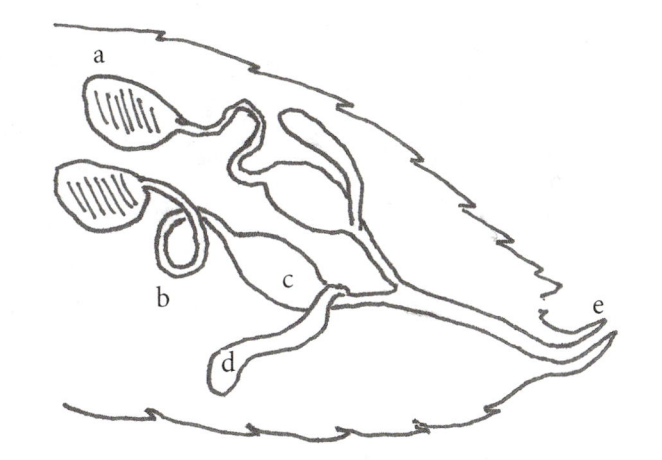

**Figure 8.1.** Schematic of a generalized male insect's reproductive tract. a. testis; b. vas deferens; c. seminal vesicle; d. accessory gland; e. aedeagus.

*Courtesy of Clyde Sorenson*

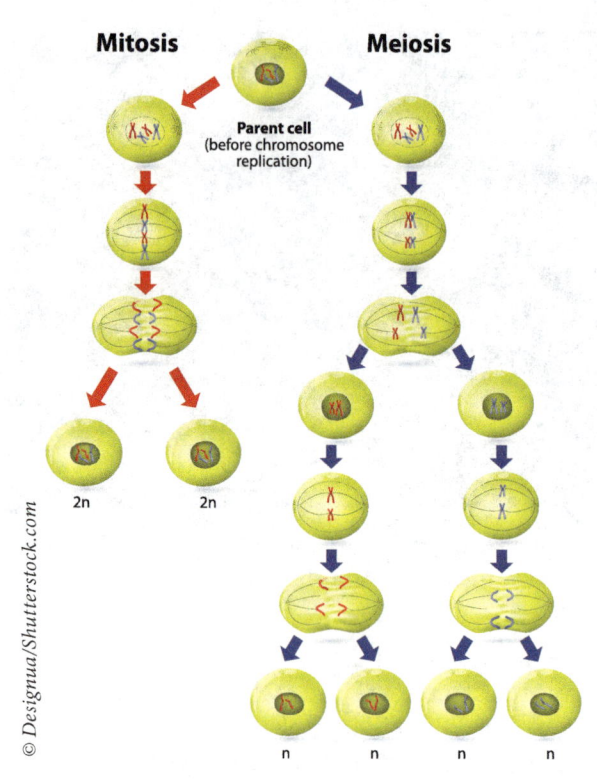

**Figure 8.2.** The difference between *mitosis* and *meiosis*.

**Figure 8.3.** The Mormon cricket, a large, wingless, gregarious, long-horned grasshopper.

primary spermatocytes undergo the two-stage process of meiosis (Fig. 8.2). In the first stage, the chromosomes replicate and may exchange information through crossover, then divide into two secondary spermatocytes. The secondary spermatocytes divide again, with each of the two new cells getting one chromosome from each of the pairs present in the secondary spermatocyte. So, each primary spermatocyte produces four sperm cells.

The testes are connected to the rest of the reproductive tract by tubes, the **vas deferens,** which transport the sperm. The vas deferens typically have an enlarged, sac-like portion called the **seminal vesicle**, where sperm are stored until needed, and they usually join into a common ejaculatory duct that leads to the external genitalia. Associated with the ejaculatory duct is a pair of **accessory glands**. Accessory glands are found in both sexes, but, of course, they do different things in each. In males, the accessory glands frequently produce substances that provide nutrition for the sperm and may also provide substances that package the sperm into packets called spermatophores.

In some species, the accessory glands also make a variety of **nuptial gifts**. Nuptial gifts are packets of nutrients or other substances the male provides at mating to provision the female as they lay the eggs that will be fertilized by his sperm. In the Mormon cricket, a large, wingless desert katydid (Fig. 8.3), the male provides a huge, protein rich nuptial gift that may equal a quarter of his body weight. In the rattlebox moth, the male's nuptial gift is a packet of toxins he harvested as a caterpillar from the host plant he fed on; the female uses the toxin to protect herself and her eggs from predators.

The sperm of insects are much like the sperm of many other animals; they have a head that contains that haploid complement of genetic material, and a tail that propels it towards the egg. While the sperm of most insects are more or the less run of the mill, the sperm of the *Drosophila* fruit flies are remarkable for their extraordinarily long tails; if stretched out, the sperm of some species are longer than the insect that made them. One species has sperm over two inches long! These are, indeed, the longest sperm in the animal kingdom.

The external genitalia of most male insects (Fig. 8.4) are composed of the aedeagus (the organ of intromission in those species—the majority—that use internal insemination, analogous to the penis in mammals) and claspers, which allow the male to retain contact with the female until insemination is completed. The shapes of the aedeagus and of the claspers are generally unique

60   Insects and People

to a species. This is one important mechanism that prevents interspecies matings. In some closely related species of insects, the only way to conclusively identify individuals visually is to microscopically examine the genitalia of the males.

## Pieces and Parts II: The Female Insect Reproductive Tract

The primary organs of reproduction in the female insect are the paired **ovaries** (Fig. 8.5). The ovaries are composed of subunits called **ovarioles**, and each ovariole is essentially a production line for eggs (insects that lay eggs in masses often lay them in multiples of the number of ovarioles in each ovary). The meiotic reduction division that produces the haploid **ova**, or eggs, happens in the tip of the ovariole, but it is a much more complicated process than that that produces sperm in the males. The eggs produced by the ovaries need to contain all the nutrients and energy the developing embryo will need to grow to hatching, so they are, of course, much larger and much more metabolically expensive to produce than sperm. As the eggs travel down the ovarioles, they accumulate nutritive yolk and, eventually, a proteinaceous "shell" called the chorion. Eggs are transported from the ovaries by tubes called **oviducts**, which typically join into a single tube called the **common oviduct**. This leads to the vagina, which receives the male's aedeagus during copulation. The vagina then leads to the external genitalia of the female, a structure called the **ovipositor**, which the female uses to place her eggs when she lays them.

**Figure 8.4.** External genitalia of a male moth.

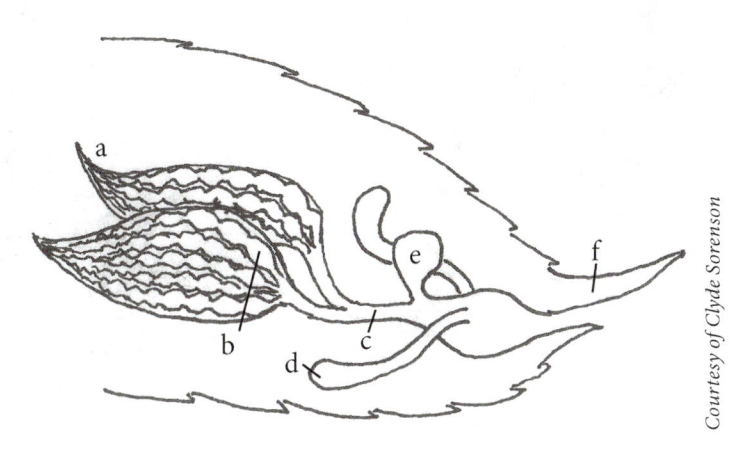

**Figure 8.5.** Schematic of a generalized female insect's reproductive system. a. ovary; b. individual ovariole within an ovary; c. oviduct; d. accessory gland; e. spermatheca; f. ovipositor.

Ovipositors come in a tremendous variety of shapes and sizes amongst insects since they do so many different things with their eggs. The long, sword-shaped ovipositor of a long-horned grasshopper (Fig. 8.6a) is used to deposit eggs deep into the soil; the saw-shaped ovipositor of a cicada is used to cut channels into twigs (Fig. 8.6b,), while the hypodermic needle-like ovipositor of a parasitic wasp is used to inject eggs into the caterpillar that will feed its larvae (Fig. 8.6c). In the workers of social hymenoptera (bees, ants, and wasps), the ovipositor ceases to have an egg-laying function (since the workers are functionally sterile), and is given over to a defensive role. It becomes the stinger these insects wield when threatened (Fig. 8.6d). If an insect stings you, you know one thing about it: it was female, since males don't have ovipositors and therefore never have stingers.

Associated with the female reproductive tract, as in males, is a pair of accessory glands Fig. (8.5). In females, the accessory glands typically produce products that help the female place and protect her eggs; in some, they manufacture a cement used to glue eggs in place, while in many Orthopteroid

**Figure 8.6.** A selection of insect ovipositors

**Figure 8.7.** A female German cockroach carrying an ootheca, or egg-case.

insects, like mantids and roaches, they produce the materials used to make the ootheca, a protective "case" for the eggs (Fig. 8.7). In stinging Hymenoptera, the accessory glands become venom glands.

Also associated with the female's common oviduct or vagina is another very important structure: the **spermatheca**. The spermatheca is a storage organ for the sperm once the insect has mated, and, in some insects, sperm can be stored here for a very long time indeed. Queen honey bees, for instance, have only one bout of mating at the beginning of their reproductive careers (but they mate with many drones over the course of those one or a few days) during which they acquire, and store in the spermatheca, all the sperm they will need for a two to seven-year mating career. The spermatheca can have a huge impact on the behavior of male insects, because, in a great many species, the last sperm that enters the spermatheca will be the first used when the female produces the next batch of eggs.

Since she is producing those massive eggs, the female's reproductive tract occupies a much larger proportion of her abdomen than does a male's; in many insects, females are substantially heavier than males because of this.

Insect eggs come in a tremendous diversity of forms and colors, but all have essentially the same structure (Fig. 8.8). The egg is contained by a **vitalline membrane** that regulates water loss, and this is surrounded by that proteinaceous **chorion**. The **nucleus** contains the genetic matter, and the

62   Insects and People

**yolk** is the energy and protein reserve required by the developing embryo. An extremely important structure on the egg is the **micropyle**—a small hole that allows access to the egg for the sperm. The micropyle passes by the opening of the spermatheca as it travels down the oviduct, and the female expresses sperm from the storage organ as it passes by.

The vast majority of insects reproduce bisexually (requiring the active participation of both a male and a female), and most insects use internal fertilization. However, there are exceptions to both rules. Some of the more ancient insects use external fertilization; an excellent example of an insect with this sort of mating system is the silverfish. In these insects, the male constructs a small silken "corral" and then deposits a spermatophore in it. He then herds the female into his corral and tries to keep her there until she accepts the spermatophore by picking it up with her genitalia, expressing the contents into her spermatheca, and then eating the empty spermatophore.

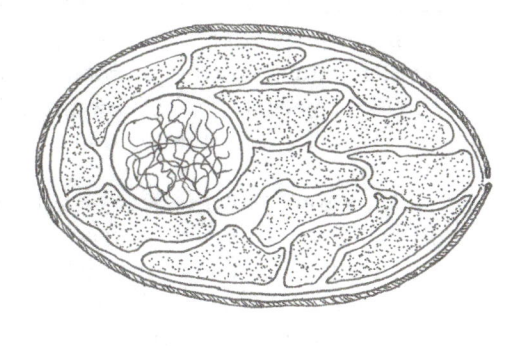

**Figure 8.8.** Schematic of an insect egg.

But again, most insects exploit internal fertilization, with the male and female coupling for some period of time while sperm are transferred, either as a spermatophore or in seminal liquid. Different insects utilize different coupling positions (Fig. 8.9), but within a species, only one position is

**Figure 8.9.** Insect mating positions.

Yes, Bees Do It: The Mating Game    63

used. In some of the more primitive hemimetabolous insects, like cockroaches, the female mounts the male, but in most higher insects the male mounts the female. Some insects remain coupled for a very long time, exceeding an hour; remember, the male is usually equipped with claspers that allow him to retain contact until he completes transferring his genetic payload. However, it should be said that in most insects, the decision to mate at all is largely up to the female.

There are some rather bizarre variations on the normal pattern of copulation. Some insects, notably the bed bugs, utilize a process called **traumatic insemination**. In this case, the male simply punctures the female's abdomen with his rather dagger-shaped aedeagus, and injects his semen into her abdominal cavity; a special structure in the abdomen receives this wound and reduces the negative impact of it. The sperm then migrate to special structures near the ovaries that collect them, and fertilization takes place in the ovaries, unlike almost all other insects. One can count how many times a female bedbug has mated by counting the number of healed mating wounds she has.

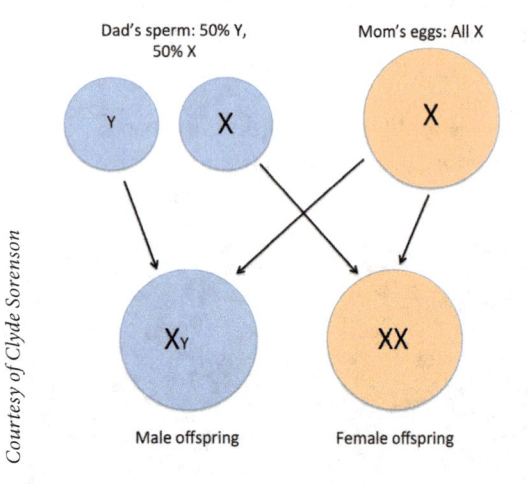

**Figure 8.10.** Typical X-Y sex determination.

Sex determination in the developing embryos of insects is generally the same as in most other animals. It is based on the sex chromosomes the embryo inherits from its parents, with X chromosomes always delivered via the egg from the mother, and either an X or a smaller Y chromosome from the father. In this system, if the embryo receives an X from both parents, it will be female, while if it receives the X from its mother and a Y from its father, it will be male (Fig 8.10). In some insects, the Y chromosome has been lost altogether; in these insects "XX" is female, while the male is "X0." (Remember, though, that they have all the other regular chromosomes. If the haploid number of chromosomes in eggs is 24, then females end up with 48 total chromosomes and males with 47.) In butterflies and moths (and birds and many reptiles), the relationship between the sex chromosomes is reversed, and males are XX, while females are XY (some folks use "W" and "Z" instead of "X" and "Y" to distinguish between the two systems). In these insects, the male always has the similar sex chromosomes, while the female has the dissimilar pair; the sex chromosome content of the egg determines the sex of the offspring rather than the sperm.

Perhaps the most significant variation on "normal" sex determination happens in the Hymenoptera: the bees, ants, and wasps, as well as a few other insects. In the Hymenoptera, a fertilized egg results in a female, while an unfertilized egg produces a male. This is called haplodiploidy; females are always diploid, males are always haploid. Females of these insects can therefore "choose" the sex of a potential offspring by deciding whether to fertilize the egg or not. This leads to a curious consequence; in colonies of social insects with one queen, who only mates one time, the sister workers are more related to each other than sisters produced through conventional sex determination. In colonies of social Hymenoptera, like ants and bees, all the workers are sterile females; the queen, of course, is also female, and only a tiny fraction of the colony's population is composed of males, whose only "job" is to mate with future queens. Popular movies consistently get this aspect of Hymenopteran biology wrong; *Antz*, *A Bug's Life*, and *The Bee Movie* all feature male workers!

Spider mites, arachnid relatives of insects, also exploit haplodiploidy. In this case, unmated, dispersing females lay unfertilized eggs that hatch into males; when these mature, they mate with their mother, and she becomes able to produce both male and female offspring.

# What Happens to the Eggs?

Most insects lay their eggs shortly after they are fertilized, usually either in, on, or very near the larval food source, and those eggs hatch some time later. These insects are said to be **oviparous**, which, basically means, "egg-laying." Birds, many reptiles, many fish, a couple mammals (the platypus and echidnas), and many non-insect invertebrates are also oviparous.

The females of some insects, like the Madagascan hissing cockroach, retain their eggs inside the vagina until they are fully mature and ready to hatch, and then appear to give birth to live young. These insects are **ovoviviparous**. Some fish, notably some sharks, and some reptiles, notably rattlesnakes, pythons, boas, and their kin, are also ovoviviparous.

A relative handful of insects are **viviparous**, meaning they give birth to live young that are nourished directly by the mother before birth, rather than relying on the egg yolk as in ovovivipary. Aphids, tsetse flies, and several other insects are viviparous, as are all mammals (except for the egg-laying platypus and echidnas).

# Asexual Reproduction—Who Needs a Mate!

In addition to more or less normal bisexual reproduction, many insects have the ability to reproduce, in one way or another, without the benefits, and problems, of mating. In all these cases, there are significant advantages, and significant drawbacks, to these strategies.

Reproducing without mating is called **parthenogenesis**, and a great many insects are parthenogenic at least some of the time. The production of males in Hymenoptera is one form of parthenogenesis; since males are the product of unfertilized eggs, they have no father and gain all their genetic information from their mother. However, the females of many other parthenogenic species produce diploid daughters without ever mating. These daughters are essentially clones of their mother, more or less genetically identical to her (except for the possibility of rare mutations). Perhaps the most familiar examples of parthenogenic species are the many species of aphids, or plant lice (Fig. 7.14). In these insects, during the parthenogenic phase of their annual cycle, females give birth to live nymphs that are the products of non-meiotic egg cell divisions in the ovaries. In other words, these females produce diploid eggs, rather than haploid eggs. Aphid annual cycles can become extremely complex, and most species do have a sexual generation at least once a year. Remember that parthenogenically produced offspring are clones of their mother, are thus genetically identical to her, and are susceptible to all the same potential threats she is.

Another form of parthenogenic reproduction, called **paedogenesis**, occurs in a handful of species of flies, beetles, and other insects. Paedogenesis means, basically, "children having children," and refers to those cases where immature forms reproduce. Great examples are the fungus midges. These insects have evolved to exploit highly ephemeral resources, like mushrooms, that don't last very long. Females lay conventional eggs on a newly discovered mushroom, and they hatch into larvae that begin feeding on the fungus. However, in order to produce even more offspring to utilize this highly temporary resource, the ovaries in the larvae develop prematurely. They produce live larvae that feed on the tissues of the mother larva and then escape through her cuticle to feed on the fungus, and perhaps go on to produce their own larvae. Some larvae may go on to develop normally and to produce pupae and then adults. Many aphids are also paedogenic, in that embryos often start developing inside nymphs before they are born; aphid mothers are therefore often carrying not only their daughters but their granddaughters as well!

A few species of insects are completely parthenogenic—males have never been found. These species may have arisen either through a mutation in a partially parthenogenic species or through the hybridization of two closely related species. While rare, this is not restricted to insects; some species of lizards are completely parthenogenic and probably arose through hybridization.

Still another case of parthenogenesis can be found in the eggs of some insects, notably some of the Hymenoptera that parasitize other insects. In these species, the adult female wasp finds and oviposits a single egg (either fertilized and female or unfertilized and male) into a host insect, and the embryo that develops from that egg divides into multiple, clonal embryos, sometimes numbering into the hundreds, that each develops into an individual larva. This is the same thing that happens in humans to produce identical twins, although much more extensive. This pattern of embryological development is called **polyembryony**. Of course, if something bad happens to that single host before the wasp larvae finish their development, all of that reproductive horsepower is lost.

## Finding That Special One—Insect Courtship

Insects have tremendous reproductive potential (more on that in a bit), but most can't exploit that potential until they find a suitable mate. Insect courtship is extraordinarily diverse and, in many cases, complex, but there are some broad ground rules. In most species, the sex that has the greatest metabolic investment in reproduction, and/or the more limited reproductive capacity, generally does the choosing, while the sex with the smaller metabolic investment in reproduction does the risky business of courting. This means that in most insects, the males court while the females assess males and chose mates. However, most insects don't live very long and therefore don't have a lot of time to wait for Mr. Right. Insects use a diversity of tactics to find each other.

Many insects use visual cues to find mates. In mayflies, the males form dense, dark swarms over their natal streams; an emerging female sees these clouds and flies up into them, where (since mayflies only live about a day as adults), the first male that sees here grabs her and hauls her out of the swarm. For males, being able to discern females is everything, so they have huge eyes compared to the females (Fig. 8.11). Male midges often form similar dense swarms over vertical structures (Fig. 8.12), which also produce strong visual cues to females.

**Figure 8.11.** Male mayfly demonstrating huge eyes.

Some other insects use stereotypical movement patterns as part of their close-range courtship communication. Several species of flies "dance" for their partners, and move their wings to display color patterns on them. Many butterflies also incorporate a dance into their courtship, often in the air (Fig. 8.13). Many beetles incorporate ritual display of horns or modified mouthparts into their courtship routines (Fig. 8.14).

Sound is extremely important in the courtship of the Orthopterans (crickets and grasshoppers), and we'll be discussing this at length in a later chapter, but many other insects also incorporate sound. Mosquitoes court with a special version of that annoying whine that sometimes keeps us awake. In these biting flies, both the male and the female call, actually harmonizing their wing sounds. Some wasps also have a song they produce with their wings to convince the females to mate with them.

A great many male insects incorporate nuptial gifts of one sort or another into their courtship behavior. Some male scorpion flies offer the female a dead insect (a highly prized food item) as a gift, and then take it back upon completion of mating to use it for another female. Others produce a proteinaceous ball of saliva as a nuptial gift. Some of the dance flies also offer a dead insect, wrapped in a silken balloon, as a gift; some species have evolved to the point that the gift is now an empty balloon!

Perhaps the most important courtship mode for most insects, however, is the chemical communication conducted through the use of **pheromones**. We will be spending a great deal more time on this in our chapter on insect communication, but for now it will suffice to say that scent is often the most efficient way for very small animals like insects to safely communicate over long distances with others of their species. Some insects employ aphrodisiacs. One can distinguish the males of the queen butterfly, and its close relative, the familiar and beautiful, monarch, by a swelling on a vein on the hind wing that the female doesn't have (Fig. 8.15); this structure is called the scent patch, and it produces a chemical that is highly arousing, and arresting, to the females. The male inserts special setae at the tip of his abdomen into the scent patch, anointing them with the chemical, then, as he overtakes her in flight, he waves them near her antennae. If she is receptive, she'll stop flying; he'll court her, and they may (or may not) mate on the ground.

Male insects often engage in heated combat over potential mates, as do the males of many other animals. Remember, if a male doesn't manage to mate during his relatively brief adult time on this planet, he is a loser in the great evolutionary game of life; his unique genetic lineage will perish with him. It is not unusual to see male butterflies engaging in spectacular dogfights, or stag beetles, with their huge, antler-like mandibles, wrestling for long minutes, over a female. It is also not unusual that some of the males of these same species will sneak in and steal the female while the top dogs battle. There are more ways then one to succeed in the mating game.

**Figure 8.12.** Male midge swarm over cemetery statue.

**Figure 8.13.** Courting swallowtail butterflies.

Before we leave this topic, we need to revisit something we mentioned earlier, and that is that the spermatheca, in many insects, determines the fate of male reproductive success, because, in many insects, the last sperm to enter the spermatheca will be the first used to fertilize any new eggs. Much of the post-mating behavior of male insects is dictated by this fact of life, and males will go to extreme lengths to ensure that their sperm do indeed produce offspring. Many males guard the female after mating, sometimes going to the extent of carrying her around until she

**Figure 8.14.** Male rhino beetles squaring off.

**Figure 8.15.** Male monarch butterfly. Notice scent patches on hind wings.

lays her eggs. In damselflies, a female can be carried off by any male that discovers her as she emerges from the water after laying eggs; male damselflies have an aedeagus shaped like a cross between a bottle brush and a spoon that they use to remove the contents of the female's genital opening before depositing their own sperm. Some other males place deposits in the female's tract that prevent other males from mating with her; in a handful, the male actually breaks his aedeagus off as a genital plug!

Before we leave the topic of reproduction, we ought to put the reproductive potential of insects into perspective, since this is one of the many important factors that contribute to the success of these remarkable creatures. Most female insects produce on the order of 100 to 300 eggs during their reproductive lives, but some quite literally produce millions of offspring. The record for a non-social insect is about 30,000 eggs. Some termite queens, though, may produce over 30,000 eggs *every day* over a 10–15 year live span.

If we then take into consideration the reproductive potential of the offspring, and their offspring, and so on, we should be knee deep in flies and neck deep in aphids. Of course, we aren't, because a great many things like to eat insects, and a great many other ills can befall them as well. Insects are so profligate in their reproductive efforts because they have to be.

# It's a Bird! It's a Plane! Insect Muscles, Locomotion, and Migration

For very small animals, insects can do remarkable things: some can fly in excess of 20 miles per hour, while others can run as fast as 3.5 mph, and still others can jump more than 100 times the length of their bodies. The remarkable locomotive abilities of insects are due, of course, to the muscles that power their limbs.

Insect muscles are, in most respects, very, very, similar to ours. They work essentially the same way as do ours, and for that matter, all animals'. Two proteins, **actin** and **myosin**, form systems of parallel fibrils that themselves are bundled together in muscle fibers; "muscles" are even larger bundles of muscle fibers (Fig. 9.1). Muscles work by contracting; muscles can only pull and never push. The contraction is achieved by the interaction of the actin and myosin. As energy is released in the muscle, the myosin repeatedly "grabs" and pulls the actin, causing the whole system to shorten. The energy source muscles use is adenosine tri-phosphate, or ATP, the fuel for most cellular work in most living things, and the release of that energy is controlled by, and precipitated by, the insect's nervous system. ATP is manufactured by the mitochondria in the muscle cells, and they are fueled by glycogen (a storage form of the sugar glucose) and oxygen, which is supplied by the tracheoles we discussed several chapters back.

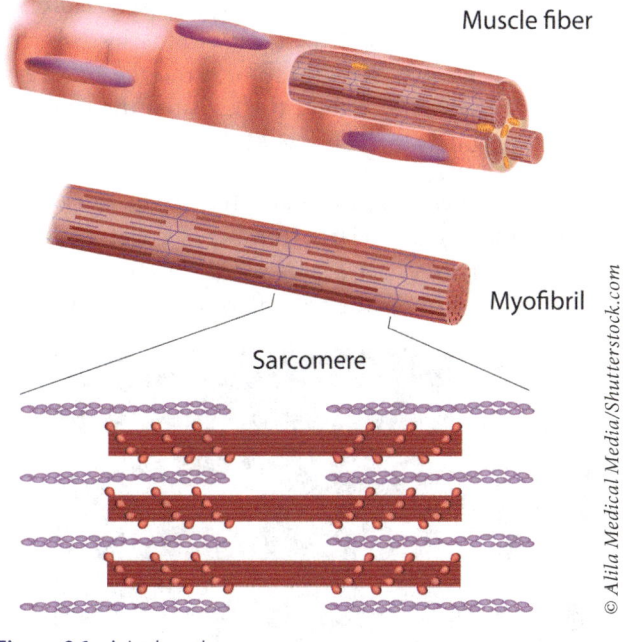

**Figure 9.1.** Animal muscle.

The ability to do work in a muscle, its strength, if you will, is dictated by its cross-sectional diameter; the bigger in diameter a muscle is, the more work it can do. Insect muscle is not inherently stronger or weaker than vertebrate muscle, although it sometimes appears that insects are incredibly strong for their size (we'll have more on this in a bit). Unlike the muscles of many vertebrates, insect muscles typically don't have iron-containing, oxygen-storing proteins like myoglobin. This together with their generally very small size, means that insect muscles generally are yellowish or whitish and translucent, more resembling the muscle meat of a shrimp than of a cow. "Red"

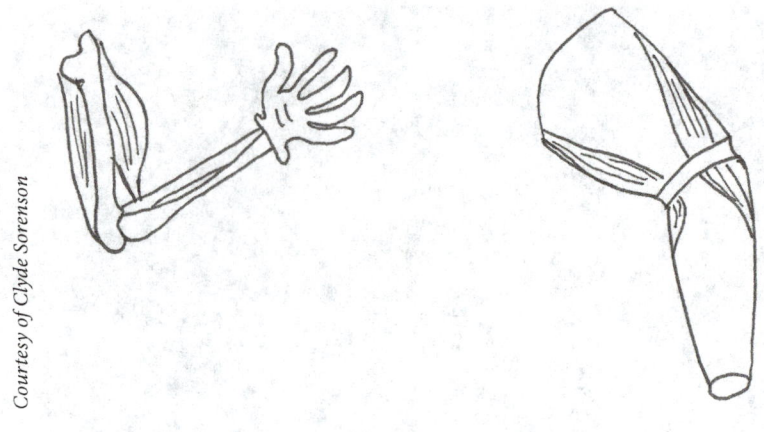

**Figure 9.2.** Opposition muscles in a human arm and an insect leg.

meat is red because of these oxygen-storing proteins. The more myoglobin a muscle contains, the darker it is, the more oxygen it can store, and the more resistant to fatigue it is. Because insect muscle doesn't store oxygen, adequate ventilation through the tracheoles is really critical to the continuous function of insect muscles. Insect muscles do fatigue, just as ours do; basically, they overrun the ability of the mitochondria to produce ATP. In response, muscles that are called upon to work for long periods of time, like the flight muscles, have much more abundant mitochondria, and so they sometimes have a slight pinkish cast.

While there are a great many variations in insect muscles, within an individual insect they can be divided into two major groups based on function: the visceral muscles, which work the internal organs, and the skeletal muscles, which move body parts. There are some differences in the ultrastructure of these two, and the muscles of the dorsal heart differ slightly from both. Visceral muscles move food through the digestive tract, eggs through the reproductive tract, and waste products through the Malphigian tubules.

All skeletal muscles are attached by at least one end to the internal surfaces of the exoskeleton. Most insects have a large number of skeletal muscles, because, thanks to their segmented body organization, they have a large number of body parts to move. Most skeletal muscles work in opposing pairs (Fig. 9.2), as they often do in us; one muscle pulls the body part in one direction, while the other pulls it back (think of your biceps and your triceps). The larger muscle in these pairs is the one that has to do the greater work; in insects with chewing mouthparts, the muscle that closes the mandible on food is much larger than the one that pulls it open again. In some insects, one of the muscles in such pairs may be replaced by a band or ball of the rubbery protein, resilin. This may save weight and may also reduce metabolic demand for the same amount of work. Many muscles and joints incorporate resilin, augmenting the force generated by the muscles.

**Figure 9.3.** A Chironomid midge.

The vast majority of insect muscles work, with respect to the nervous system, as do our skeletal muscles. One nerve impulse results in one contraction. We call such muscles **synchronous**—the muscle activity is synchronized with the activity of the nervous system. Some, however, respond differently to a single nerve impulse. They contract multiple times for each nerve impulse, resulting in much higher muscle contraction rates- as much as 10 times higher. These **asynchronous muscles** are found powering the wings of many holometabolous insects and in the thorax of some male cicadas, powering the tymbals they use to make their loud courtship

70   Insects and People

songs. Asynchronous muscles tend to be elastic, and they oscillate in concert with the organ they move; the act of being stretched at the peak of an oscillation predisposes them to contract for the next. Asynchronous muscles enable some insects, certain species of midges (Fig. 9.3), to flap their wings 1,000 times per second, or 60,000 times per minute!

Insects do appear to be abnormally strong, but this is mostly an illusion related to their small size. The strength of muscles, as we stated above, is dictated by their cross-sectional diameter; strength increases in muscles as the square of this area. But body mass is related to volume, and volume increases as the cube of length. Because of the repercussions of this "square-cube" law, if a one-gram grasshopper, for instance, can jump a meter high, then a 100-gram grasshopper won't be able to jump 100 meters. It'll most likely still only be able to jump about a meter high! Of course, we've already determined that a 100-gram grasshopper is unlikely due to the limitations of an exoskeleton and tracheal respiration. It must be said, however, that an exoskeleton does provide for extremely efficient muscle attachment, and that can enhance the efficiency of muscles. In grasshoppers, for instance, the large leg muscles are attached obliquely to the exoskeleton, in effect greatly increasing their cross sectional diameter and therefore strength. Other adaptations can also enhance the efficiency of insect muscles; fleas can jump to such incredible heights, in proportion to their bodies, by storing large amounts of energy in elastic resilin pads associated with their jumping legs. It's much like slowly compressing a large spring, and then releasing it to effect a jump; their legs function as levers, increasing the efficiency.

While muscles are important in maintaining almost all life functions, their most prominent and obvious function is locomotion. Insects have a tremendous variety of locomotion modes they can exploit. They can, of course, walk, run, and fly, but they can also swim and use various modes of crawling. We'll now explore some of these modes.

# Walking, Running, and Crawling

We previously discussed the different types of legs insects have: the cursorial, saltatorial, natatorial, raptorial, and fossorial variations on the basic leg. In all these cases, the activity of the legs on a particular thoracic segment is controlled by the nerve **ganglion** in that segment (Fig. 6.14). A ganglion is a cluster of nerve cell bodies. Relatively few neurons control the muscles in an insect leg, but the thoracic ganglia and these few neurons coordinate the movements of the legs. This is why a headless insect can continue to walk in a more or less normal (if somewhat disturbing) fashion.

Six legs allow insects to take advantage of an efficient and stable form of locomotion. They walk by shifting alternating tripods composed of the front and hind leg on one side together with the middle leg on the other side (Fig. 9.4). Tripods are extremely stable structures, so insects using this strategy are always very solidly supported by the substrate they are walking on. While this seems like a relatively straight-forward process, it actually demands a very high degree of coordination, since each of the legs in each tripod actually has to do something slightly different in order to shift forward, plant itself on the substrate, and pull the insect forward.

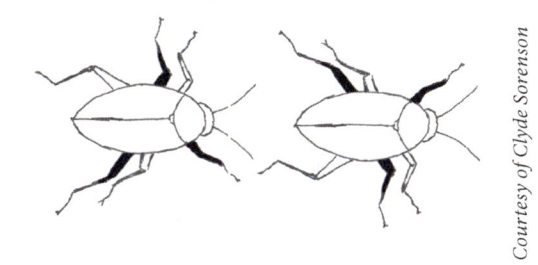

**Figure 9.4.** Tripod locomotion.

**Figure 9.5.** American cockroach.

It's a Bird! It's a Plane! Insect Muscles, Locomotion, and Migration

**Figure 9.6.** Green June beetle larva and adult.

**Figure 9.7.** Palm weevil larva.

**Figure 9.8.** Flat-headed wood-boring beetle larva.

**Figure 9.9.** Cecropia caterpillar with crochets on prolegs.

When an insect runs, this order breaks down somewhat, and only one leg may be in contact with the ground at any given time. No animal can run with more frequent steps than the fastest insect runners (cockroaches); an American cockroach (Fig. 9.5) can take 20 steps a second! Proportional to their size, cockroaches may be the fastest runners on the planet, even if their top end speed in real terms is only about 3.5 mph (although the ones that used to skitter across the kitchen floor in my old grad school apartment seemed much faster).

Larvae of insects that go through complete metamorphosis exploit a tremendous diversity of ways to move their typically soft, long bodies around. Most beetle larvae that have legs (many don't) basically use the same walking strategy that adult insects do, and simply drag their abdomens around. The larvae of the green June beetle (Fig. 9.6), however, flip over onto their backs and use a vertical undulation to crawl. Many of the legless beetle larvae, like those of weevils and tree borers (Fig. 9.7 and 9.8), live in galleries they create by chewing their way through the substrate. These insects can move by alternately expanding and contracting different parts of the body to gain a purchase on the sides of the gallery and then pull themselves forward or backward.

Most caterpillars have, in addition to their six legs, unjointed appendages called prolegs (Fig. 9.9) on some of their abdominal segments, and these are armed with tiny, curved hooks called crochets. A crawling caterpillar grasps the twig or leaf it is on with all six legs and the prolegs, and moves by reaching forward, from front to back, with first its "real" legs and then each set of prolegs, so that it always firmly grips its substrate with most of its appendages. The caterpillar we call an inchworm (Fig. 9.10) has only two sets of prolegs located at the end of its abdomen, and crawls by reaching forward with its six legs, grasping the twig, and then releasing the prolegs, arching the body and regrasping the twig, appearing to measure the twig in the process.

Fly larvae, or maggots, are legless, and many live in semiliquid environments. Maggots either move through serpentine wriggling movements or by the same waves of contraction we described above for legless beetle larvae. Some also appear to use their mouth hooks to assist their movement. Mosquito larvae, of course, live in water and swim by vigorous serpentine wriggling.

**Figure 9.10.** A Geometrid caterpillar called an "inchworm."

# Wings and Flight

The most exceptional and remarkable locomotive ability insects possess, of course, is their ability to fly. There is certainly no better way for a small animal to cover vast distances. Flight allows insects to exploit habitats and ecological niches other invertebrates can't. While there is still a great deal of debate over the evolutionary origin of wings, we do have a pretty solid understanding of how they function.

Insects have wings only on the mesothoracic and/or metathoracic segments. All the muscles that power the wings of insects are located in the thoracic box, as are the tiny muscles they use to change the aspect (angle) as they navigate. The wings are essentially double-layered plates of cuticle reinforced by a system of thickened veins. More primitive insects, like dragonflies (Fig. 9.11) tend to have many more veins and cross veins than more advanced insects like bees and flies (Fig. 9.12). The arrangement of veins can be a very useful tool in classifying insects. The veins are also important in expanding the insect's wings at its last molt; hemolymph is pumped into the collapsed veins before the cuticle tans. The major veins also contain a nerve and trachea to provide oxygen to the living tissues in the wing.

**Figure 9.11.** Carolina saddlebags dragonfly.

The muscles located in the insect's thorax that power flight may, in some insects, constitute 40% of the animal's weight. In the most primitive flying insects, the dragonflies and mayflies, the major flight muscles are attached directly to the bases of the wings (Fig. 9.13), and the wings are caused to move up and down by alternate contractions of sets of these muscles. Insects with these **direct flight muscles** are able to move the hind wings independently of the forewings, and so you may see the forewings at the top of the flapping stroke while the hind wings are at the bottom.

**Figure 9.12.** Golden snipe fly.

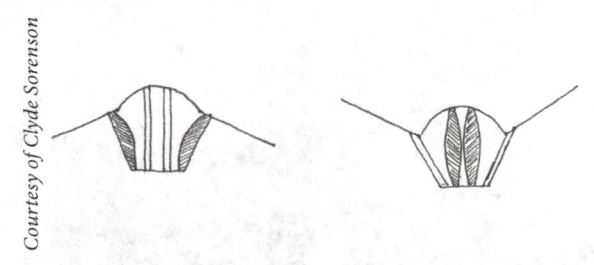

**Figure 9.13.** Schematic of direct flight muscles.

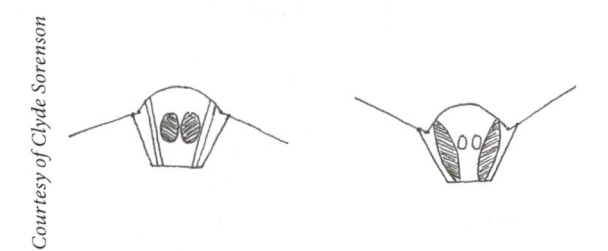

**Figure 9.14.** Schematic of indirect flight muscles.

**Figure 9.15.** The wings of a white ibis. Note bones visible through extended wing.

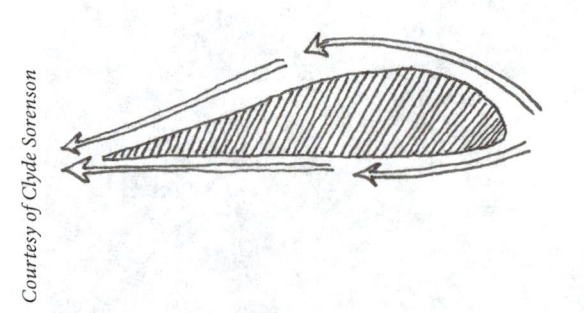

**Figure 9.16.** Schematic of the cross-section of an airfoil.

Most flying insects, however, have **indirect flight muscles** (Fig. 9.14). In these insects, there is no direct connection between the major flight muscles and the wings. The muscles instead are attached to the walls of the thorax, and they cause the wings to go up and down by deforming the thoracic box. One group of muscles links the top and bottom of the thorax, while the other extends lengthwise through the thorax. The wings are linked to the top (notum) and sides (pleura) of the thorax in a way that forms a toggle joint. When the vertical muscles contract, the roof of the thorax is pulled down and the wings pop up; when the longitudinal muscles contract, the roof of the thorax pops up and the wings are forced down. In these insects, the forewings and the hind wings move in concert because both pairs are responding to the deformation of the thorax. In many, the forewings are linked to the hind wings by a mechanism that essentially turns two wings into one, as we discussed in Chapter 5.

The wings of insects and the wings of birds are similar in some functional respects but very different in others. The wing of a bird is the highly modified forelimb of a vertebrate quadruped (Fig. 9.15); the inner part of the wing is composed of the upper and lower arm bones (humerus, radius, and ulna), while the outer part is the hand. The inner part supports the tertial and secondary feathers, and in most birds, these and the other feathers that clothe this part of the wing form an airfoil (Fig. 9.16), while the outer part supports the primaries, which are sort of like a propeller—they pull the wing forward through the air. An airfoil provides lift through its shape; air has to travel further over the top of the wing than across the bottom, and this reduces the air pressure above the structure. The inner part of a bird's wing, therefore, provides lift, while the outer part provides propulsion. As anyone who has ever enjoyed a bucket of Buffalo wings can testify, there are abundant (and delicious!) muscles in the wing of a bird, and a bird's wing in flight is constantly changing shape as the creature navigates its way through the air.

An insect wing, on the other hand, is a more or less flat plane when it's at rest, at least at first glance. There are no muscles outboard of the thorax, so there is apparently no way the wing's shape can be dramatically changed. An insect wing therefore must provide both propulsion and lift simultaneously, and so insects fly in a rather dramatically different way from birds. Most bird flight relies on a steady flow of air over the

74   Insects and People

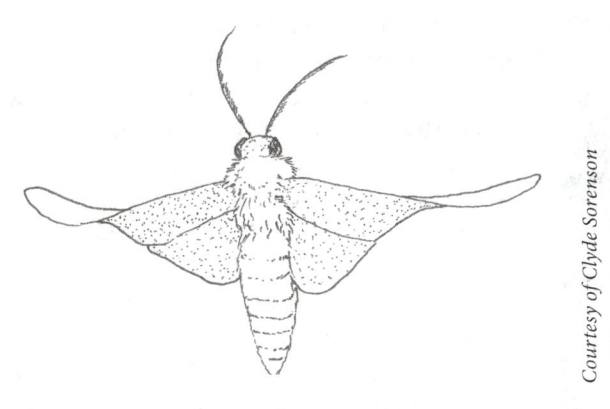

**Figure 9.17.** A moth in flight demonstrating the steep angle of attack and deformation of the wing.

airfoils; insect flight instead relies on the unstable air flow they produce with their wings. The wings don't just go up and down; they also are rotated dramatically at the apex of both the up stroke and the down stroke (Fig. 9.17) to produce a very high angle of attack. This causes ring-shaped, rolling "leading edge" vortices, which provide lift and are pushed back and down. This kind of flight only works at the high wing beat frequencies most insects use, since the lift provided by the vortices is very transient. Hummingbirds fly much more like larger insects than they do other birds, and indeed, the flight of a hummingbird and of a hawk moth are remarkably similar (Fig. 9.18). Insects and hummingbirds both can hover, and their skill at this remarkable behavior is again due to their ability to create and exploit zones of unstable vortices (Fig. 9.19), except in this case, the vortices move in a vertical direction. The wings of a large, gliding insect like a monarch butterfly, on the other hand, interact with the stable air flow more like a conventional bird's wing does when it sails on motionless wings.

**Figure 9.18.** (a) Pink-spotted Hawk Moth; (b) Rufous hummingbird.

**Figure 9.19.** Unstable vertical vortices on the hawk moth in figure 9.18a.

# Migration

The impressive locomotion abilities of insects allow them to efficiently do one thing that less mobile animals can't do very well—they can migrate long distances. Animals migrate to avoid predictable, inhospitable environmental conditions, or to exploit concentrations of valuable resources like food. While most migratory insects migrate by flying, some insects can move fairly remarkable distances on foot. Insects exhibit at least three types of migratory movement patterns: **Nomadic migration**, **one-way seasonal migration**, and **two-way seasonal migration**.

*Nomadic migration.* Nomadic migration is perhaps best typified by the various species of migratory locusts. Locusts are large grasshoppers, and the migratory species are generally found in tropical and sub-tropical, arid and semi-arid parts of the world where grasslands predominate. Most of the time, members of these species are solitary animals that are most interested in eating grass and

**Figure 9.20.** Gregarious form nymph of the desert locust.

**Figure 9.21.** Adult desert locust.

finding the occasional mate. They actively avoid each other except when courting and mating. These "solitary phase" locusts are generally well camouflaged by their drab coloration, and while they can fly quite well, they have relatively limited flight range. Under the right environmental conditions, however, local locust populations start to increase as breeding success increases, and eventually, individual locust nymphs start encountering other members of their species with increasing frequency. This forced contact causes a change in both the behavior and the physiology of the insects, and instead of trying to avoid each other, they start to seek each other out. They become brightly colored (Fig. 9.20), and start moving about in large, dense bands. The adults that emerge from these roving mobs of nymphs are more brightly colored than solitary adults and are larger with longer and more powerful wings (Fig. 9.21). These are the insects that can form swarms numbering into the billions of individuals. Swarms move in response to local food depletion. Swarms can move hundreds of miles in a day, depending on the speed of the wind, since the swarms almost always move downwind. The direction of migration, then, isn't determined by the locusts themselves, but by the wind direction, but, since winds usually move in the direction of low-pressure weather systems, and low-pressure systems often spawn rains, this movement pattern often steers the swarm in the direction of fresh, new food resources, where they may also find conditions suitable for breeding. Locust swarms can cover thousands of miles over several months; swarms of the desert locust of North Africa have traveled as far as the Arabian Peninsula, southern Europe, and even across the Atlantic to the Caribbean Islands!

Locust swarms cause untold harm to human populations in much of Africa and, to a lesser extent, parts of Asia since they can cause the virtual destruction of crops in a matter of days. Very little can be done to alleviate the damage done when a swarm descends, and, unfortunately, the areas most likely to experience locust swarms are also often home to very poor people who rely on subsistence agriculture for their livelihoods. The international community is expending a great deal of effort to figure out ways to deal with these insects.

Nomadic swarming migration is not restricted to insects that can fly, by the way. The Mormon cricket (Fig. 8.3), the large wingless katydid we discussed earlier and common across much of the Great Basin of western North America, can build to huge populations under conditions similar to those that initiate locust swarms. Mormon cricket bands migrate by walking, and they, too, can be devastating to any agriculture they encounter. It's been discovered that Mormon crickets within a band keep walking to avoid being eaten by the crickets moving behind them. Migrating crickets crave protein and salt, and their fellow band-mates can be an excellent source of these nutrients.

During one outbreak that occurred in Nevada while I lived there, Interstate 80 was temporarily closed for a time because the road had become slick with the crushed bodies of a band of crickets.

Mormon crickets acquired their somewhat erroneous name when they threatened the first crops of the early Mormon settlers in Utah. The swarming bands were wiped out by flocks of California gulls from the nearby Great Salt Lake; a statue dedicated to the birds that saved the first settlers still stands in Salt Lake City, although some historians dispute the veracity of the tale.

*One-way seasonal migration.* Some insects make predictable annual migrations. In the northern hemisphere, these migratory movements are often from south to north in the spring. The insects that exhibit this pattern are frequently species that can't tolerate the rigorous winter conditions of higher latitudes, but can persist in more tropical regions. A good example of such an insect is the potato leafhopper a small hemipteran insect that causes big problems for agriculture. The leafhopper survives the winter in the relatively moderate temperatures along the Gulf Coast and in the southern Coastal Plain, living on evergreens and herbs that grow in these areas. In the spring, the warming temperatures cause the insects to leave en masse; steering currents associated with the jet stream transport them north, spreading the insects to northern and eastern states, where they can cause significant damage in crops like alfalfa and peanuts. In the fall, some of the insects return south to the Gulf region, but most die. Like the leafhoppers, most insects that exploit this kind of migration are moving in response to inhospitable environmental conditions or limiting resources, like food. An important distinction is that, even when there are both northward movements in the spring and southward movements in the fall, the same insects are not going in both directions, and where individual migrating insects wind up is often due largely to the steering weather patterns.

*Two-way seasonal migration.* Some insects engage in annual migrations in which at least some individuals travel both north and south in a directed fashion. Probably the most famous of these insects is the monarch butterfly (Fig. 9.22). Monarchs are not able to cope with freezing temperatures. This poses no problem in the tropical areas they inhabit, but it is, of course, a significant issue in most of the United States and Canada. Monarchs solve this problem by migrating south, beginning in August. What sets monarchs apart from almost all other insects is that virtually all the monarchs in eastern North America find their way to a very small area high in the mountains of central Mexico, where they aggregate in their millions in a handful of groves of fir trees (Fig. 9.23). These groves offer the

**Figure 9.22.** Monarch butterflies.

perfect environment to allow the butterflies to survive the winter; they are cool enough to prevent the insects from exhausting their fat reserves before winter's end, but they rarely see any frost. They also offer the butterflies the perfect humidity, and most groves have small streams where they can drink. The butterflies are in reproductive diapause over the winter; they spend their time clustered in the trees, and on warm days, they fly down to springs and streams to drink. In March, the reproductive diapause breaks, and the butterflies frantically mate (an interesting story in its own right) before departing to the north. The females lay eggs on the milkweed host plants as they go, but most are able to only travel a few hundred miles northward before they exhaust their resources and die. Their offspring complete their development in about five or so weeks, and they continue migrating northward, laying eggs as they go. By the third generation, monarchs have once again spread throughout eastern North America and have reached, once again, southern Canada. Their late-summer offspring, once they complete their development, fly to the south, and the annual cycle begins again. The butterflies that overwinter in Mexico may live as much as nine months, three times longer than any of the summer generations.

How the insects find their way back to this isolated and very tiny group of groves (the total area of all the groves together is less than 50 acres in most years) has long been a source of mystery. We have learned that monarchs use a sun compass and a magnetic compass, both located in their antennae, to help them navigate. They get in the general vicinity of the groves by maintaining a southwestward heading as the fly. Once they arrive in the Transvolcanic Belt Mountains in Mexico, the first arrivals rely on their ability to smell the odors left by the millions of last year's butterflies in the fir needles of the previously used trees. They continue to fly, at the right altitudes (9,000-11,000 ft.), until they find the trees that smell right. Later arrivals also use the orange color of earlier arrivals to help them find the right trees. The flocks of butterflies may shift up and down the mountain as weather conditions demand. By the way, the monarchs from the western parts of North America have a similar migration pattern. These overwintering populations concentrate in central coastal California, although the numbers of butterflies and the size of the aggregations is typically much smaller than in the Mexican refuges.

**Figure 9.23.** Monarch butterfly aggregation in oyamel firs in Mexico.

The winter aggregation of the monarchs is one of the most spectacular natural phenomena in the world, but it is endangered. For many years, the wintering groves have been threatened by illegal logging, although this seems to have abated somewhat in recent years. However, the populations of the butterflies have continued to decline precipitously. More recent declines have been attributed to a series of severe droughts in central North America, coupled with extensive use of herbicides that kill the milkweeds larvae need in agricultural landscapes. Concerted action is required to protect this grandest of insect displays.

# Bugs, Bees, Beetles, and Butterflies: The Diversity of Insects

Insects are clearly the most diverse creatures on the planet, and they have been around for a very, very long time. Their great time on this planet means that they aren't simply diverse in terms of species, but that they are also extremely diverse at higher taxonomic levels. Insects have exploited virtually every available habitat on land and in freshwater, and, between this and their great age, it is extremely difficult to sort out their evolutionary history. For most of the history of the science of insect systematics, the science of describing the evolutionary history of the creatures, we have only had tools of observation. We could study their appearance and structure, and we could watch them as they developed, but these technologies only allowed us to see but so deep into their past. Over the last 25 years, however, our understanding of genetics (built largely on deep study of one insect—the fruit fly *Drosophila melanogaster*) has improved to the point that we can now read the "permanent record" of evolution housed inside every cell of every living thing, and this has revolutionized our understanding.

The most recognizable higher taxonomic level for most people is the order—these are the large groups that most people can readily recognize. The number of orders of insects has varied, and continues to change. It has ranged between 24 and 36 orders in the last several decades, but most current systematists put the number at about 30 (Fig. 10.1). There are an additional three orders of closely related, very primitive, non-insect hexapods (they used to be considered insects, but some important differences we'll discuss below have changed that). Insects within an order always undergo the same kind of metamorphosis, and they generally have the same kinds of mouthparts and wings. Most also have some unique distinguishing characters, although sometimes they can be difficult to see. The orders fall into important

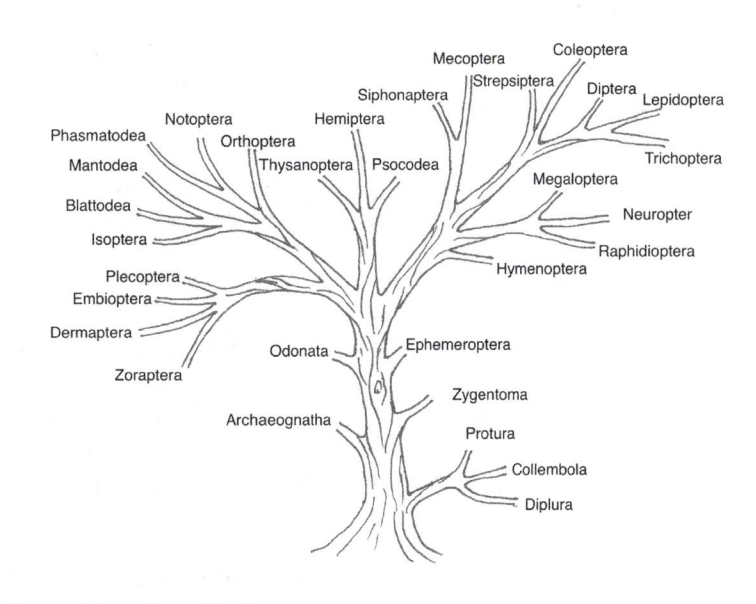

**Figure 10.1.** A "Tree of Life" for insects—one of many.

groupings that also inform our understanding of evolution. We are going to briefly survey the orders of hexapods below, starting with the most ancient.

# The Non-insect Hexapods

The non-insect hexapods are all wingless, and they share in common an odd and interesting arrangement of their mouthparts, which are tucked into a hood on the head; we call this condition **entognathy** ("hidden mouth"). All these creatures are very small, and all primarily inhabit the soil. There are three orders of these entognathous hexapods. Like the most primitive true insects, these creatures undergo ametabolous lifecycles; during mating, all exploit external fertilization.

**Figure 10.2.** A slide-mounted Proturan. This animal is about 3 mm long.

*Protura.* The Proturans (Fig. 10.2) are perhaps the most ancient living hexapods; Protura means "first-tailed." They are distinguished from all others by several unique characters: they have 12 abdominal segments (All others have 11), and no antennae. Instead, they use their first pair of legs much like antennae. They also have rudimentary appendages on the first three abdominal segments. Protura are soil dwellers that feed on fungi. There are about 500 species of Protura, but, unless you make a concerted effort, you are not likely to ever see one.

*Collembola.* The Collembola (Fig. 10.3) are also called springtails, and, unlike the Protura, they are extraordinarily abundant in some places and very easy to find. Springtails are called "springtails" because of a pair of interesting structures on the abdomen of most of the 6,000 or so species. A lever-like structure called a furculum is located on the fourth abdominal segment. It latches into a "catch" on the third called a retinaculum. A startled springtail releases the catch, and the furculum flings it into the air, and, hopefully, away from the hazard that startled it. Springtails also have a structure

**Figure 10.3.** A springtail.

called a collophore on the underside of the first abdominal segment (this is where the order name comes from—Collembola means "glue-wedge"). This structure appears to secrete a sticky substance and may help them regulate body water. Most Collembola are extremely small animals; the largest are only a quarter inch long, and most are much smaller. Collembola have three or four immature instars and continue to molt after reaching adulthood.

Collembola live in a huge diversity of habitats, including grasslands, marshes, seashores, tree canopies, and, most significantly, soils. They can reach astounding densities; a single square meter of lawn could have thousands of the creatures at some times of the year. Heavy rainfall often flushes large numbers out of turf. After such an event, the surface of rain puddles may have an "oil sheen" composed of springtails, which hop on the surface of the water like popcorn in response to a hand waved over them. Most springtails feed on fungi and detritus, and a couple are pests for mushroom growers. They are ecologically very important as primary decomposers of plant matter and as prey for other invertebrates.

*Diplura.* The Diplura (Fig. 10.4) are another small order (about 800 species world-wide) of small and inconspicuous animals. They vaguely resemble the Protura, except that they do have antennae, and they also have well-developed cerci. In some, the cerci are forceps-like, much like those of an earwig, and these species use the cerci to capture prey. Diplura live in leaf litter and soil; some are fungi and detritus-feeders, while those with the forceps-like cerci are aggressive predators. Like the Protura, they often have rudimentary appendages called styli on some abdominal segments. Unlike Protura, Diplura continue to molt after reaching adulthood.

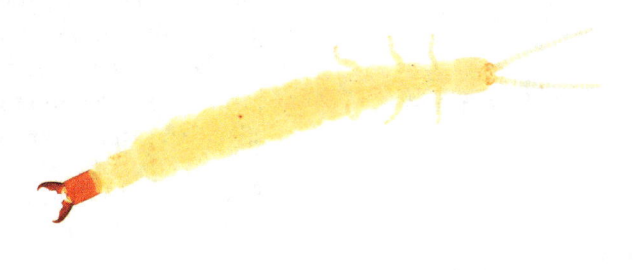

**Figure 10.4.** A slide-mounted Dipluran. This animal is about 5 mm long; you are looking at its ventral side.

## The Primitively Wingless Insects

All insects are ectognathous, meaning that their mouthparts are exposed. There are two orders of insects that are wingless and that evolved from wingless ancestors; all other wingless insects evolved from winged ancestors (there will be more on this later). These two orders are probably the most ancient living insects, and they share some significant traits in addition to their primitively wingless condition. Insects in both these orders practice external sperm transfer, the bodies of both are usually clothed in scales that shed easily, and both continue to molt after reaching sexual adulthood, unlike all other true insects. These two orders resemble each other enough that they have long been considered one order, the **Thysanura**, but a great deal of work over the last 20 years or so has conclusively proven that they are two orders.

*Archaeognatha.* The Archaeognatha are also called bristletails (Fig. 10.5) or jumping bristletails. There are about 500 species worldwide. These are by far the most ancient living insects; they have been around for close to 400 million years. They have many primitive characteristics, including styli on their abdominal segments. They have three tail appendages and large eyes. Bristletails have very primitive chewing mouthparts and feed primarily on fungi and algae. They live in many different habitats, but most places they live tend to be humid. One of the most remarkable things about the bristletails is their ability to jump; rather than using their legs to propel themselves into the air, they forcefully flex their abdomens against the substrate much like a crayfish or lobster uses its tail to propel it forcefully backwards through the water it lives in. Compared to the silverfish in the order Zygentoma, the bristletails in North America tend to be browner in color, have a "humpbacked" body shape, and their three tail appendages tend to all be directed backwards.

*Zygentoma.* The Zygentoma are the silverfish and firebrats (Fig. 10.6). There are about 500 species of these insects. While they look much like the bristletails, they are actually much more closely related to the flying insects we'll discuss shortly. Silverfish tend to be dorso-ventrally flattened and tapered towards the tail. They have three tail appendages, and the two cerci are typically held at almost a right angle from the body. They are called silverfish because most have bodies clothed in silver scales. This, with their wiggly, rapid running gate

**Figure 10.5.** A jumping bristletail.

Bugs, Bees, Beetles, and Butterflies: The Diversity of Insects

make them look like little, wriggling minnows. Most have small compound eyes or no eyes, and feed on all sorts of organic matter. Like the bristletails, the silverfish continue to molt, probably to replenish the protective scales and to get rid of infective agents on the cuticle. Some silverfish can live for five or six years and may molt 40 or more times!

Silverfish are minor pests in and around homes, and can occasionally damage books and other paper products. A few live as "impostors" in ant nests. Typically, ants kill any foreigners that invade their nests, but these silverfish chemically disguise themselves so that the ants don't recognize them as different.

**Figure 10.6.** A silverfish.

# The Ancient Wings—Paleoptera

The two most primitive living orders of flying insects share some important characteristics that differentiate them from all other flying insects, and together they are known as the Paleoptera, or "ancient-wings." The most significant of these characters is that these insects do not have the ability to rotate their wings and fold them down over their abdomens. This is because they don't have the more complicated articulations between the wings and the thorax that more advanced flying insects possess (Fig. 10.7). As a result, these insects either hold their wings straight out or vertically over their backs when at rest. Both of these orders have aquatic immature stages.

*Ephemeroptera.* The Ephemeroptera are the mayflies (Fig. 10.8). The order's name refers to the extremely brief adult life of these insects. Most only live for a day or two as adults, and in one, the American sand-burrowing mayfly, *Dolania americana*, the adult females only live for about five minutes! Mayflies are quite distinctive. They are delicate insects with long, soft abdomens, usually three long caudal (tail) appendages, and triangular, gossamer wings with a great many veins, which they hold upright over their backs. The only duties of the adults, during their very brief lifespan, are to mate and disperse; therefore, no mayflies feed as adults, so they have vestigial mouthparts. There are about 2,500 species of mayflies worldwide.

**Figure 10.7.** Paleopteran wings (a) vs. Neopteran wings (b).

Uniquely in the entire sweep of insect diversity, mayflies are the only insects that molt functional wings. When the last nymphal instar is completed, the nymph swims to the surface of the water and rapidly sheds its nymphal exuvium. What emerges is a stage called the **sub-imago**—a soft and drab creature with translucent wings (Fig. 10.9). It flies weakly to a nearby perch and almost immediately molts again. This time, what emerges is a glistening, fully colored animal with glittering, clear wings, capable of much more adept flight and of mating. This **imago** may rest for a short time, but generally pursues an opportunity to mate very soon since its time is so short!

All mayflies are aquatic as nymphs, and they, too are pretty distinctive (Fig. 10.10). Most mayfly nymphs have three tail filaments, like the adults, and most have obvious gills on the sides of their abdomen.

**Figure 10.8.** A Mayfly.

82  Insects and People

They occupy a tremendous diversity of ecological niches, but most live in cold, clean, well-oxygenated water. Most are detritivores (eating small bits of decomposing organic matter), or grazers on algae, but a few are predaceous on other insects. Mayflies are critically important in freshwater ecosystems. They play an important role in nutrient recycling and are extremely important as food for many other aquatic animals, particularly fish like trout. Consequently, many humans who are otherwise disinterested in insects have great affection for and knowledge about mayflies and spend much time creating artificial, hooked versions of them out of feathers and bits of floss that they use to capture the finny denizens of their favorite stream. (I suffer from this affliction myself.) Since they require clean, well-oxygenated water, the diversity and abundance of mayfly nymphs in a body of water can be an extremely effective measure of the quality and environmental integrity of the stream or lake.

*Odonata.* The Odonata are quite familiar to most folks. These are the dragonflies and the damselflies (Fig. 10.11 and 10.12). The name Odonata refers to an odd arrangement of their mouthparts. It seems to me a better name for such a remarkable and conspicuous order could have been found, but there it is. The Odonates are all predators as both immatures and adults, and they are all aquatic as immatures. There are about 6,000 species worldwide. The dragonflies and the damselflies fall into to two suborders; the dragonflies are the Anisoptera, while the damsellies are the Zygoptera. There are some conspicuous differences between these two groups.

Damselflies are usually the smaller and more delicately built Odonates. At rest, most fold their wings vertically over their backs, much like mayflies. In most, the forewings are roughly the same size and shape. The adults tend to fly low and slow, weaving their way among stalks of grass and the branches of shrubs, looking for tiny flies and other prey while trying to avoid being eaten themselves. The aquatic nymphs of damselflies typically have three leaf-like external gills at the apex of their abdomen, and they, too tend to be long and delicately built creatures (Fig. 10.13). While most are denizens of still water habitats like lakes and ponds, some are restricted to fast flowing streams. While our North American species are at most only a couple of inches long as adults, some tropical species are as long as five inches, with seven-inch wingspans, although they still have the same delicate build.

Dragonflies are typically much more robust than damselflies, and, in general, are powerful and agile in the air. Most dragonflies hold their wings horizontally, or even angled slightly downward, when perched. The hind wings and forewings of dragonflies differ in shape; in most the hind wings are broader. Dragonflies are familiar and almost ubiquitous in fresh and even brackish water habitats. Some species are adapted to life on fast flowing streams, while others exploit ponds, lakes, and marshes. Dragonfly nymphs

**Figure 10.9.** A mayfly sub-imago. These are the only insects in the world that molt wings.

**Figure 10.10.** A mayfly nymph.

**Figure 10.11.** A banner clubtail dragonfly.

**Figure 10.12.** A sparkling jewelwing damselfly.

**Figure 10.13.** A damselfly nymph (naiad).

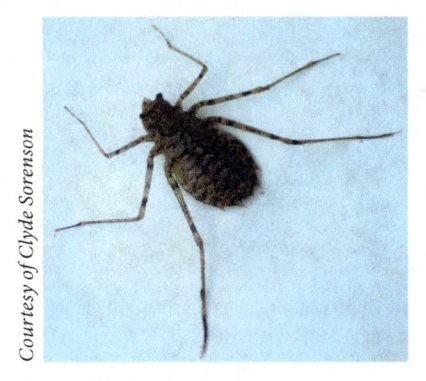

**Figure 10.14.** A dragonfly nymph (naiad).

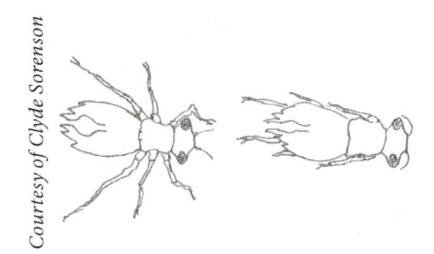

**Figure 10.15.** Schematic demonstrating anal jet propulsion of a dragonfly nymph.

**Figure 10.16.** Halloween pennant dragonflies in the "wheel" mating position.

(Fig. 10.14) have no external gills; rather, they pump water in and out of their rectums, which are surrounded by a dense network of trachea. This unusual method of respiration also gives dragonfly nymphs a rather unusual method of escape. They can use jet propulsion, by forcefully expelling water from their rears, to shoot them through the water (Fig. 10.15).

All Odonate nymphs are predaceous, and most will capture and eat anything small enough to subdue. Some of the bigger dragonfly nymphs regularly feed on small fish and tadpoles, along with the aquatic insects they more typically capture. Odonate nymphs have unique mouthparts to facilitate their hunting. Their labium is expanded forward into an elbowed "mask" that covers the lower part of the head at rest (Fig. 4.11). The palps at the end are expanded, shovel-shaped structures armed with sharp teeth. When prey approaches close enough, the labial mask whips out, much like a snake striking, and the nymph grabs the creature with the labial palps, then pulls the prey into the waiting mandibles. The strike of a dragonfly nymph is one of the fastest movements in the animal kingdom.

Odonates also have rather unusual mating behaviors, due to the unique arrangement of the sex organs of the male. The testes and most other bits of the male's reproductive system are located near the end of the abdomen, as in most insects, but the copulatory structure is located on the underside of the second abdominal segment, near the thorax. When ready to mate, the male transfers sperm to this structure, and then, upon finding a willing female, grasps her behind the head with the claspers on the end of the abdomen. The female must bend her abdomen forward to engage the copulatory organ (Fig. 10.16).

## The "New" Wings—The Neoptera

All other orders of insects are Neopteran—meaning that they can rotate their wings and fold them over their abdomens. Even though some of these orders are now wingless, they evolved from Neopteran ancestors. We call this condition "secondarily wingless" as opposed to the primitively wingless insects we discussed earlier. Neopterans have the ability to fold their wings because they have evolved additional sclerites and muscles near the base of the wings. Being able to rotate and fold their wings gives these insects a tremendous advantage over the Paleopterans because it gets the wings out of the way as these insects crawl about through the underbrush. All Neopterans share a common ancestor.

Neoptera are divided into several major groups.

## The Stoneflies and their Allies

All four of these orders are hemimetabolous and have chewing mouthparts.

*Plecoptera.* The order Plecoptera is composed of about 3,500 species of stoneflies, vaguely roach-like insects with aquatic nymphs. The flattened adults (Fig. 10.17) have relatively narrow membranous forewings and huge, fan-shaped hind wings. They usually have well-developed cerci, and chewing mouthparts. The nymphs (Fig. 10.18) also have well-developed cerci and gills usually located at the junction of the legs and the thorax. Stonefly nymphs, like mayfly nymphs, demand clean, well-oxygenated water and are thus good bioindicators of water quality. They are also extremely important food for trout and other vertebrates inhabiting the same waters. Some stoneflies are rather exceptional in that the adults have their mating flights on mild days in the depths of winter.

*Embioptera.* The Embioptera is a small order of generally very small insects that look a great deal like little stoneflies (Fig. 10.19). All are terrestrial and live colonially in leaf litter or tree bark, where they eat lichens and fungi with their chewing mouthparts. They are sometimes called webspinners because they construct silken tunnels with glands located on their forelegs. They can run just as fast backwards as forwards, which is a useful adaptation to living in their silken galleries. Only the males are winged. There are about 400 described species of these relatively poorly known insects.

*Zoraptera.* Zoraptera is a very small (39 identified species) order of very poorly known, tiny insects (Fig. 10.20). They are vaguely termite-like in appearance and in behavior. They live in small colonies under bark or logs and feed on fungi. Most are wingless (Zoraptera means "Pure wingless"), but some do have wings. You are extremely unlikely to ever encounter these creatures unless you make a concerted effort to do so.

*Dermaptera.* The Dermaptera, or earwigs (Fig. 5.20), are probably the most familiar of these four orders since a few species are minor pests in houses and greenhouses. The name "earwig" is an ancient mistake, based on a long-standing misapprehension that these creatures actually like inhabiting human ears. (Many insects have been retrieved from human ears over the ages, but very few are earwigs.) Earwigs are quite distinct insects thanks to one obvious characteristic—the large, forceps-like cerci at the back end. Earwigs use these defensively, and predaceous species may use them to help capture prey. In many species, they are larger in the male than the female and may have some role in sexual selection. Earwigs also have short, leathery forewings that protect larger, fan-shaped, membranous hind wings. Earwigs may be plant feeders or predators on other insects, and most are most active at night; a few are parasitic. Earwigs are subsocial, meaning that the female does guard and care for her eggs and small nymphs. Many also deploy a foul-smelling liquid as a defensive strategy when threatened.

**Figure 10.17.** Adult stonefly.

**Figure 10.18.** A stonefly nymph.

**Figure 10.19.** A webspinner of the order Embioptera.

**Figure 10.20.** A zorapteran.

**Figure 10.21.** A short-horned grasshopper.

**Figure 10.22.** The Jerusalem cricket, a long-horned "grasshopper."

**Figure 10.23.** Oriental cockroaches.

**Figure 10.24.** Termites of the order Isoptera.

# The Polyneoptera

All these orders are hemimetabolous and have chewing mouthparts and hind wings (if they have hindwings) large and fan-like. We once referred to these as the "Orthopteroid" orders because these are the most familiar members of this large and diverse group.

*Orthoptera.* The Orthoptera are the grasshoppers and crickets. All have saltatorial, or jumping, hind legs, although they are much better developed in some than others. Orthopterans fall into two main groups. The diurnal Caelifera (grasshoppers and locusts) (Fig. 10.21) have short antennae and short or non-existent ovipositors. The mostly nocturnal Ensifera (katydids and crickets) (Fig. 10.22) generally have long, thread-like antennae and conspicuous, and often long, ovipositors (Ensifera means "sword-bearer"). The Ensifera are probably the more diverse, including both the burrowing mole crickets and the arboreal katydids. There are about 20,000 species of Orthopterans. Orthoptera means "straight wings," and refers to the narrow, leathery forewings, called tegmina, that protect the large, fan-shaped, membranous hind wings that do most of the work in flight.

Most Orthopterans are plant feeders, and some are important agricultural pests. They can be extraordinarily abundant and several species, as we discussed previously, are notorious for swarming, nomadic migrations. They are also probably the most widely consumed as food by humans. A giant weta from New Zealand is a contender for the heaviest adult insect in the world at about 3 oz., heavier than many mice!

Orthoptera are also noteworthy for their use of stridulatory sound in their courtship communication. Caelifera tend to have their stridulation organs on their legs and wings or abdomen, while Ensifera stridulate with their wings exclusively. Ensiferan "ears" are on their forelegs, while Caeliferan "ears" are on the abdomen.

*Blattodea.* The Blattodea (sometimes called Blatteria) are the cockroaches (Fig. 10.23). Cockroaches are probably the most reviled of all insects, because of their association with filth, but only about 15-20 of the approximately 4,500 species are pestiferous. Most of the rest are tropical forest dwellers that are important players in nutrient recycling. We'll spend some time on the pest species in a later chapter. Cockroaches are typically dorso-ventrally flattened and have tegmina protecting their hindwings. Most have a large, shield-shaped pronotum that protects the base of the wings and the cervical ("neck") area behind the head so that you typically can't actually see the head when viewing a roach from above. Most have long antennae and well-developed cerci, and many are very swift runners; many also fly quite well. Cockroaches range in size from less than a quarter of an inch in adult length to almost three inches. In many species, the females carry an ootheca around for quite some time before depositing it. Some others, like the Madagascan hissing cockroach, are ovoviviparous, and retain their eggs until hatching. Some primitive species are subsocial. In these species, a pair will raise and protect a relatively small cohort of offspring until they are nearly

86   Insects and People

mature. Cockroaches are very closely related to the termites; indeed, most systematists now consider termites to be eusocial cockroaches that specialize in digesting cellulose.

*Isoptera.* The Isoptera are the termites (Fig 10.24). While, as I just said, the termites appear to have evolved from cockroaches, they are distinct enough and important enough to recognize on their own. Termites are the most prominent hemimetabolous eusocial insects. Eusociality is thought to have evolved in these insects due to the difficulty in utilizing their cellulose-rich diet. Most termites are relatively small, soft-bodied, white insects, and when you find one, you find many. Termites have well-developed caste systems. Typically only one female, the queen, produces all the eggs for the colony, with one male reproductive called the king (in some species, however, there are secondary queens and kings). Workers and soldiers maintain and defend the colony. Termites can't digest cellulose without the symbiotic assistance of bacteria or protozoa that inhabit their guts. In most species, these gut symbionts have to be regained from another termite after every molt. The name Isoptera means "same wings"—the deciduous forewings and hind wings of alate reproductives (future kings and queens) are similar in size and shape.

**Figure 10.25.** The aardvark, the ultimate end for many a termite.

While termites can be very important structural pests in our dwellings and other structures, their ecological significance is hard to understate. They are often the primary recyclers of dead plant matter in many ecosystems, and a great many other animals rely on them for food, sometimes exclusively. The aardvark (Fig. 10.25) specializes in termites, as do the aardwolf and "anteaters." Termites also generate copious amounts of the greenhouse gas, methane. There are about 4,000 species of termites.

**Figure 10.26.** A Chinese mantid eating the first American copper butterfly I ever saw in my yard.

*Mantodea.* The Mantodea are the mantises (Fig. 10.26). Mantises are very closely related to roaches and termites, and, indeed, are grouped with them in the supraorder Dictyoptera. Mantises are functionally predaceous roaches in this context. All 2,500 species of mantises are aggressive predators on other animals, and all are armed with distinctive, raptortial forelegs at the front of an equally distinctive, elongated prothoracic segment. Most will eat any animal they can capture and subdue, including small vertebrates like lizards, frogs, and even hummingbirds! Mantises exploit camouflage to protect themselves from their many predators but also to disguise themselves from their potential prey; some even mimic flowers to good effect (Fig. 10.27).

**Figure 10.27.** A flower-mimicking orchid mantis.

Mantises, like cockroaches, typically protect their eggs in an ootheca. In mantises, however, this structure is a made of a frothy, styrofoam-like substance glued to a twig (Fig. 10.28).

*Phasmatodea.* These are the walking sticks (Fig. 10.29). Most are elongate, twiggy-looking creatures that rely on that appearance to camouflage them from potential predators. Some tropical and subtropical species, however, look more like living or dead leaves (Fig. 10.30), sometimes down to simulated insect damage! All are plant feeders. There are about 3,000 species. Most stick insects lay their eggs singly, often simply dropping them to the forest floor, and some may take over a year to reach adulthood. Some species also reproduce at least partially through parthenogenesis. In addition to camouflage, some species use noxious chemicals to defend themselves, and many also have startle patterns on their hind wings that can be used to intimidate predators. The longest insect at present is a recently discovered stick insect from Southeast Asia. With legs extended, this creature is over 20 inches long!

Bugs, Bees, Beetles, and Butterflies: The Diversity of Insects

**Figure 10.28.** A Carolina mantis ovipositing in a mason jar on my living room mantle.

**Figure 10.29.** A walking stick insect.

**Figure 10.30.** A leaf-mimicking stick insect.

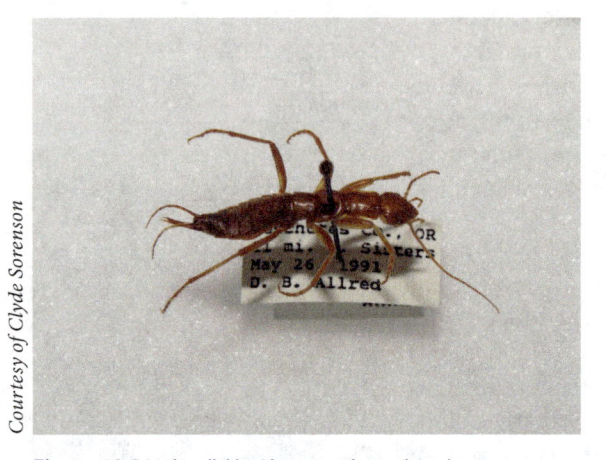

**Figure 10.31.** A grylloblattid snow roach or rockcrawler.

**Figure 10.32.** The only Mantopasmatodid "gladiator" specimen in the NCSU entomology museum.

*Notoptera.* The insects in this extremely small (only about 50 species) order have some of the most bizarre lifestyles among the insects. All are wingless. There are two suborders within this odd order. The first contains the rockcrawlers (Fig. 10.31), small, roach-like insects that live at high altitudes, often near the foot of glaciers or permanent snowfields, and feed on the carcasses of other insects paralyzed by the cold. Unlike almost all other insects, they are most active at temperatures just under 40° F.

The other suborder was only discovered in the last 20 years or so in southern Africa; these are the gladiators or heelwalkers (Fig. 10.32). These insects were known as fossils in amber, but a worker noticed one in a collection of pinned specimens, and this prompted exploration in the area where that specimen was collected. All of these creatures are predators on other insects. They look something like a cross between a mantis and a stick insect, but their first two pairs of legs are raptorial! These insects are restricted to semi-arid areas along the southwest coast of Africa.

# The Paraneoptera

These orders undergo hemimetabolous lifecycles, don't have cerci, and most have mouthparts adapted to feeding on liquids.

*Thysanoptera*. The 5,000 or so species of thrips compose the order Thysanoptera. Thrips (Fig. 10.33) are generally very small insects (most are only a couple millimeters long) with elongate bodies, rather cone-shaped heads, and asymmetrical mouthparts adapted for probing individual plant cells and sucking out the liquids within. They have one functional mandible, usually the left one, that punctures the cell. The right side maxilla is formed into a tube for removing fluids. Some are predaceous, primarily on other thrips. Thrips have odd wings that are essentially a rod fringed with long setae, so that each looks like a tiny bird feather. (Thysanoptera means "fringed-wing.")

While thrips are hemimetabolous, their lifecycle is a bit more complicated than that of most other insects with simple metamorphosis. The last immature stage is inactive, doesn't feed, and often spins a cocoon, but the creature within has wing buds and the same body conformation as previous immature instars. Thrips have haplo-diploid sex determination; in most species, unfertilized eggs produce males and fertilized eggs females, but, in some, the reverse occurs. Some thrips form colonies with some of the characteristics of eusociality, probably due at least in part to haplo-diploidy.

Some thrips are serious pests on crop plants, both because of the direct damage they do to buds, flowers, and leaves, but also because they transmit plant diseases. Some of the predaceous species are important natural enemies of pest thrips and spider mites.

*Hemiptera*. The Hemiptera are the true bugs. They all have piercing-sucking mouthparts, but they are very diverse in other characters. There are three major groups of Hemipterans, and these are distinct enough that they were once recognized as two separate orders. The "true" true bugs, or Heteroptera, have their mouthparts originating from the front of the head, scent glands on their thorax, and hemelytral front wings (the kind that forms an "X" pattern when folded over the abdomen). Perhaps the best example of this group is the stink bug (Fig. 10.34). Other examples include assassin bugs, giant water bugs, and bed bugs. Heteroptera have very diverse feeding patterns. Some are exclusively plant feeders (and are important agricultural pests), some are exclusively predators, some feed on both plants and animals, and some are parasitic on vertebrates. Some of the parasitic species are important disease vectors. Chagas disease, an important and increasing tropical disease, is transmitted by a group of parasitic assassin bugs called cone-noses. There are about 25,000 species in this group.

The second group is the sub-order Sternorryncha. These are the aphids, scale insects, and their relatives (Fig. 10.35). Sternorryncha have mouthparts that originate at the rear of the head. If they have wings, they are usually membranous and held tented over the abdomen at rest. Most of these insects are small, soft bodied, and inconspicuous until they reach really

**Figure 10.33.** A thrips.

**Figure 10.34.** A brown stink bug.

high population levels. Many are sessile as adults, meaning that once they settle on their host plant, they never move again. Many protect their bodies with waxy scales or waxy flocking produced by special glands in their cuticles. All are exclusively plant feeders and many are extremely important crop pests. They can cause direct injury by sucking the nutrients the plant has made for itself, but also can vector serious plant diseases. Some have an economic value of another sort—both dyes and shellac can be manufactured from some scales. There are about 12,000 species in this group.

**Figure 10.35.** An alate (winged) greenbug, a type of aphid.

The third group is the sub-order Auchenorryncha, which contains the various tree- and planthoppers, and cicadas (Fig. 10.36). Like the Sternorryncha, these insects have mouthparts that originate at the back of the head (these two sub-orders were once lumped together in the order Homoptera), and they have membranous wings held tented over the abdomen. However, unlike the Sternorryncha, these insects are often brightly colored, very active, and generally (but not always!) larger. All are plant feeders, and, again, some are important agricultural pests and plant disease vectors. Many species in this group have sub-social parental care. The longest lifecycles, egg to egg, in the insects are also found in this group in the form of the periodical cicadas of eastern North America. Three of these species require 17 years to complete their lifecycle. There are about 30,000 species of hoppers and their kin.

*Psocodea.* The Pscodea is an interesting order because, for a very long time, the three primary groups in it were considered orders in their own right and their relationships were not very well understood at all. This order contains the bark lice and true lice, and there are about 8,000 species.

Bark lice and booklice (Fig. 10.37) are small, soft-bodied insects with mouthparts somewhat intermediate between sucking and chewing organization. They feed by scraping material with their mandibles. If they have wings, they are membranous and held tented over the abdomen. The wings often have patterns of dark marks. Bark lice are generally found, as the name suggests, under bark or other dead wood, where they feed on fungi and lichens. Many are colonial, and the nymphs are often brightly marked. Some species are parthenogenic.

**Figure 10.36.** A treehopper from the North Carolina mountains.

Book lice are generally minute, wingless, whitish insects that feed on starchy materials. These are the tiny white things you sometimes see (no, you didn't imagine it) crawling across the pages of an old book in the library. Book lice are rarer in libraries than they once were because most libraries curate their collections, and control the storage environment for their books much more effectively than even a couple decades ago.

The true lice were once known as the Phthiraptera. All lice are wingless parasites that live on vertebrates.

**Figure 10.37.** A colony of bark lice.

The sucking lice (Fig. 10.38) were once recognized as the order Anoplura. All sucking lice are parasites on mammals and feed on their hosts by piercing their skins with their mouthparts and imbibing the host's blood. Most are very intimately tied to their specific hosts and can't live long away from them. In fact, the relationships between lice have been used to help clarify the relationships between the host animals. Sucking lice usually have dorso-ventrally flattened bodies and legs adapted for holding on to the host's hair. The eggs of lice are called nits and are typically glued to the hair of the host animal. All nymphal stages and both adult sexes rely solely on the host for their sustenance. As might be expected, sucking lice can be important vectors of diseases, and some are serious livestock pests in their own right.

The chewing lice (Fig. 10.39) were once considered to belong to the order Mallophaga. The vast majority of chewing lice are parasites on birds, although a few have mammalian hosts. Most feed on the dander and surface skin cells of the host animal. The primary negative effect on the host is usually the irritation resulting from the scraping activity of the insects' mouths on the host skin, but some feed aggressively enough to draw blood. Like sucking lice, species of chewing lice have very tight affinities with specific host species and generally can't survive away from them. A few species are important livestock pests. The chewing lice are evolutionarily more diverse than the sucking lice, and their taxonomy is, therefore, more complicated.

**Figure 10.38.** Human crab or pubic louse.

**Figure 10.39.** Three bird lice.

# The Holometabola—The Evolutionary Success of Complete Metamorphosis

The remaining orders are all holometabolous, meaning they undergo complete metamorphosis. Over 80% of all insect species fall into this group, which suggests that, at least in terms of species diversity, there is a great benefit to this kind of life cycle. Holometaboli allows insects to efficiently occupy different habitats as immatures and adults, which allows them to exploit much smaller niches. There are about 11 orders of holometabolous insects.

*Hymenoptera.* The Hymenoptera are the familiar bees, ants, and wasps, and the less familiar sawflies. The order derives its name from the way the fore- and hind wings are linked together; in these insects, the leading edge of the hind wing has a row of tiny hooks (Fig. 5.16) that engage the trailing edge of the forewing. The order's name has an interesting and convoluted origin. *Hymeno* refers to the joined membranes of the fore- and hind wings. Hymen was the ancient Roman god of the wedding, and his presence was necessary for a happy marriage; thus the membrane found in the human female's genital tract that is (often erroneously in both directions) thought to indicate virginity, ended up with Hymen's name. The larvae of these insects are typically legless (except for the sawflies) and have a well-developed head capsule with chewing mouthparts. Most adults have chewing mouthparts as well, though some have the maxillae and labium modified for drinking nectar. Hymenoptera have haplo-diploid sex determination; fertilized eggs become female, while unfertilized eggs produce males. There are about 150,000 species.

**Figure 10.40.** A sawfly. This is a tropical species that feeds on cocoloba.

**Figure 10.41.** Pine sawfly larvae.

The sawflies (Fig. 10.40) and their near relatives are the most primitive Hymenopterans; unlike all others, they have a thick junction

Bugs, Bees, Beetles, and Butterflies: The Diversity of Insects

**Figure 10.42.** A cicada-killer wasp.

**Figure 10.43.** A leaf-cutter bee.

**Figure 10.44.** An adult antlion, a neuropteran relative of lacewings.

**Figure 10.45.** Green lacewing larva.

**Figure 10.46.** Racket-tailed neuropteran.

between the thorax and abdomen, rather than the "wasp-waist" of wasps, ants, and bees. Sawfly females often have serrated ovipositors they use to insert their eggs into plant stems, giving the group its name. Sawfly larvae (Fig. 10.41) look very much like true caterpillars except that they have unjointed prolegs on all abdominal segments, and these lack the little hooks, called crochets, that help caterpillars grip the substrate they're crawling on. Sawfly larvae are also unusual for Hymenoptera in that many are foliage-feeders or stem borers, and some are important agricultural and forestry pests.

The remaining Hymenopterans all share a common ancestor. The vast majority of these species are commonly called wasps (Fig. 10.42). Wasps fall into a large number of families and range in size from incredibly small (the smallest flying insect is a tiny parasitic wasp) to more than two inches long. Most wasps either hunt other insects as prey, or parasitize them; some are scavengers. The ovipositor of the parasitic forms is usually a modified stinger the insect uses to insert her egg(s) into the host.

Bees are essentially wasps that have shifted from a diet of insect protein to a diet of nectar and pollen harvested from flowers. Bees (Fig. 10.43) are generally "hairier" than wasps; these setae help the insect to collect pollen as it visits flowers. Many bees convert nectar to honey by treating it with enzymes from their guts and dehydrating it. They range from solitary to eusocial. Bees are, of course, critically important to the pollination of many plants, including a huge range of crop plants.

Ants are essentially highly social (eusocial) wingless wasps. All ants belong to the single family Formicidae, and all are eusocial. Ants have distinctive elbowed antennae and most are fairly small; they may or may not have stingers. Different species of ants feed on a huge range of food. Some are highly predaceous, while others specialize in seeds and still others specialize in feeding on one type of fungus. Ants also demonstrate a tremendous range of higher social behaviors, including tending livestock, slave-making, and gardening. Ants, like the termites we discussed previously, often dominate the ecosystems they live in, both in numbers and in biomass, compared to other animals. Some ants are important pests of agriculture and in the home, and some ants, bees, and wasps are, of course, medically significant.

*Neuroptera.* The Neuroptera are the lacewings, antlions, mantispids, and their allies (Fig. 10.44). All are predaceous or parasitic as larvae, and most adults are predaceous as well. The adults are rather delicate-looking creatures with fairly soft bodies, chewing mouthparts, and large, membranous wings with a great many veins and cross veins (Neuroptera means "nerve-winged," and refers to these abundant veins). The larvae are usually somewhat squat creatures with remarkable and unique sucking mandibles (Fig. 10.45). Most capture prey by plunging their sickle-shaped "jaws" into the hapless victim and sucking it dry. Larval Neuropterans have a remarkable diversity of predaceous lifestyles; some are aquatic and graze on freshwater sponges, while others scurry about plants looking for aphids. Still others (the antlions) dig cone-shaped traps and wait patiently at the bottom, while a few are parasitic on spiders and Hymenoptera. Some of the oddest and most intriguing insects belong to this order (Fig. 10.46). There are about 6,000 species of Neuropterans world-wide.

*Megaloptera.* The Megaloptera are the 300 or so species of dobsonflies and fishflies. These animals are obviously fairly closely related to Neuropterans; they share the same large, heavily veined wings and soft bodies. However, the Megaloptera are typically larger, and sometimes much larger, and all have aquatic larvae, called hellgrammites, that vigorously prey on other animals with their chewing mouthparts. The adult males of the dobsonflies are huge and imposing, with great sickle-shaped mandibles crossing out in front of their heads. Females lack these, suggesting their role in sexual selection. The fishflies or alderflies are smaller and much less intimidating in appearance (Fig. 10.47). Hellgrammites (Fig. 10.48), in addition to being important predators and prey in the streams they inhabit, make excellent fish bait. They also bite, so they should be handled with care.

*Raphidioptera.* The Raphidioptera are the snakeflies (Fig. 10.49). Snakeflies also superficially resemble Neuropterans, but the adults have elongated prothoracic segments that allow them to rear up rather like a cobra (but with far less devastating consequence). The 200 or so species of snakeflies are all terrestrial/ arboreal predators both as adults and as larvae, and they are pretty much restricted to areas with cold winters. In North America, they are restricted to mountainous areas of the West.

*Coleoptera.* The Coleoptera are the beetles, the most mega of the four megadiverse orders of insects (Coleoptera, Hymenoptera, Lepidoptera, and Diptera). There are approximately 350,000 described species of beetles, which dwarfs all other taxa. (Although it must be admitted that Hymenoptera and Diptera may prove to be as, or more, diverse than beetles if ever we approach describing all of them.) The hallmark of the beetle, and probably one of the main reasons for the success of this incredible assemblage of creatures is the elyrtra—the often hard, protective, forewings of the animal (Fig. 10.50). These structures allow beetles to inhabit virtually any kind of substrate the planet can offer and still maintain the power of flight. They can tunnel through wood, burrow through dung, bore through plant stems, and paddle through water, and then take to the air to find the next such patch of habitat. As a result, beetles are pretty much everywhere.

Beetles have chewing mouthparts both as adults and as larvae, although these may be modified a great deal in some species to accommodate unique diets. Most beetle larvae are soft-bodied, with a well-developed head capsule and short legs; weevil larvae, however, are legless, and wireworms, the larvae of click beetles, have tough, heavily sclerotized exoskeletons (Fig. 10.51). The largest proportion of beetles probably feeds on plants in one way or another, but a significant minority is predaceous at some stage, and a few are parasitic on other arthropods as larvae.

**Figure 10.47.** A fishfly.

**Figure 10.48.** A hellgrammite, the aquatic larva of a Dobson fly.

**Figure 10.49.** A snakefly.

**Figure 10.50.** A longhorned beetle.

Beetle diversity is actually too overwhelming to even begin to summarize here. Let it suffice to say that, if it can be done by an insect, there is probably a beetle doing it. They range in size from virtually microscopic to approaching five inches in length; the larva of a goliath beetle is another contender for the heaviest living insect, at about four ounces. Not surprisingly, Coleoptera contains a large number of economically important insects, including beneficial predators, agricultural pests, pollinators, dung recyclers, and forestry pests. Some of the latter are currently restructuring the forests of the Rocky Mountains in response to climate change.

*Strepsiptera.* The Strepisptera are a small order of very odd, tiny, insect parasitoids; their closest relatives appear to be the beetles. The common name for these insects is the "twisted-winged parasites," but this is really only suitable for the adult males since the females are wingless larviform parasites. First instar larvae are mobile and seek a host insect. Once they find an appropriate host, they enter its body and molt to a legless, grub-like creature. Females pupate within the host body, and, upon becoming an adult, exert a portion of their abdomen through the host's cuticle. This bit often resembles a tick emerging between the abdominal sclerites of the host (Fig. 10.52). Males pupate and then emerge as a free-living, flying form that seeks the female in her host. After mating, the female produces a new batch of free-living first instars that leave her body to find their own host. Males are the only flying, two-winged insects in which the hind wings are the functional flight organs, and the forewings have evolved into flight-stabilizing halteres (Fig. 10.53). They also have odd, raspberry-like compound eyes. There are about 600 species of Strepsipterans in the world; they parasitize Hymenoptera, Orthoptera, and Hemiptera.

*Mecoptera.* The Mecoptera are the scorpionflies. Scorpionflies (Fig. 10.54) are called this not because of any venomous hazard they present, but because the males of some species have abdomens and terminal sexual apparatus that resembles the curled "tails" of scorpions. The order name means "long-wings," and refers to the long, narrow, and often patterned forewings. The 600 or so species of scorpionflies fall into several families, including the typical scorpionflies, the snow scorpionflies, and the hanging scorpionflies. To some extent, the scorpionflies are the vultures of the insect world, since many species feed primarily on insect carrion. The head of most species is extended into a long proboscis, with the chewing mouthparts at the end. This arrangement allows the insect to chew through the cuticle of a dead insect and then feed on the soft tissues within by probing around in this small hole, much as a turkey vulture might feed on the muscle of a dead deer by inserting its head and long, naked neck into the carcass through a small tear in the hide. Some species are predaceous.

Scorpionflies are somewhat notorious for the complicated mating rituals many species perform. They make extensive use of nuptial gifts. In some species, the male makes a ball of protein-rich, rubbery saliva, while in others, the gift is a dead insect. In most, the size of the nuptial gift dictates the duration of mating, and so males strive to present the

**Figure 10.51.** Larva of the eyed elater, a large click beetle.

**Figure 10.52.** A *Polistes* wasp with a female strepsipteran parasite on its abdomen.

**Figure 10.53.** A male Strepsipteran. Note the forked antennae and "raspberry" eyes; the knob-shaped structures in front of the wings are its forewing halteres.

**Figure 10.54.** A scorpion fly.

largest gift they can obtain. Some rob spider webs for suitable gifts, and some end up paying with their lives in the venture.

The larvae of scorpionflies are somewhat caterpillar-like and feed, like their parents on other insects, dead or alive.

*Siphonaptera*. These are the fleas. Siphonaptera means "wingless sucking pipe," more or less, and that's a pretty apt description of a flea. All fleas (Fig. 10.55) are wingless parasites on vertebrates, primarily mammals, although some do feed on birds. Fleas are secondarily wingless insects—they evolved from winged ancestors, in this case, from scorpionflies. They are laterally flattened, to facilitate rapid crawling through fur, and have numerous "combs" of stout spines that make it more difficult for the host to remove them through grooming. They also have saltitorial hind legs, assisted by deposits of resilin in the exoskeleton, that give them the remarkable jumping ability for which they are famous. The larvae of most species are legless, maggot-like creatures that feed primarily on the fecal pellets of adult fleas and on detritus. The adult females of some fleas attach to the host for prolonged periods of over week, but most fleas hop on and off the host animal as their hunger demands, spending the bulk of their time in the den or bedding the host uses.

**Figure 10.55.** A cat flea.

Adult fleas feed with piercing mouthparts powered, in part, by bands of elastic resilin. Since they potentially feed on multiple hosts, they can be important vectors of disease in man and animals. As we'll see later, tiny fleas have had a tremendous impact on the course of human history.

*Diptera*. The Diptera are the true flies; these insects have two functional mesothoracic wings, while the hind wings are reduced to flight stabilizing halteres. The Diptera (Fig. 10.56) is the second or third largest order in terms of described species, with about 120,000 identified thus far. However, there may be two or three times more yet to be described. The flies are probably the most ecologically diverse insects. The order includes predators, parasitoids, plant feeders, saprophages, vertebrate parasites, aquatic forms, and terrestrial forms. All flies are adapted to eating either liquid foods or foods that can be liquefied prior to ingestion. They may accomplish this with piercing-sucking mouthparts, sponging mouthparts, or slashing blades used to liberate blood from a vertebrate. The order is divided into two major groups: the more primitive Nematocera, including crane flies, mosquitoes and their kin, and the more advanced Brachycera, which includes horseflies, houseflies, and the like.

The larvae of flies are typically legless and with poorly defined head capsules (Fig. 10.57), although Nematoceran larvae do have recognizable heads. Most live in aquatic, semi-aquatic, or extremely damp environments. Most larvae feed as saprophage—scavengers that live off the dead remains of other living organisms, ranging from disintegrating leaves to the corpses of animals. However, there are some that graze on algae, and some that feed on living plants. Many are parasitoids on other insects or parasites on vertebrates.

**Figure 10.56.** A robberfly.

Flies are another group, like the beetles, that exhibit diversity far greater than we can begin to be described here. They are perhaps the most skilled and acrobatic fliers of all insects. Horseflies are among the fastest insects, and no other animals achieve the wing beat frequencies (up to 1,000 beats/ second) of some of the midges.

**Figure 10.57.** Fly larvae, or maggots.

Bugs, Bees, Beetles, and Butterflies: The Diversity of Insects

**Figure 10.58.** California gull feeding on brine flies at Mono Lake.

**Figure 10.59.** Lake Malawi with swarms of lake flies in the distance.

**Figure 10.60.** Soldier butterfly.

**Figure 10.61.** Luna moth caterpillar.

**Figure 10.62.** An eastern tent caterpillar moth.

They may live in more extreme environments than most—hypersaline Mono Lake in California grows tremendous populations of brine flies (Fig. 10.58), and the natural crude oil seeps of La Brea support populations of petroleum flies that actually flourish in the otherwise toxic hydrocarbons. Flies may achieve some of the greatest densities, in some places, of any terrestrial animal on earth (Fig. 10.59). They are, of course, critically important economically and medically. Diptera includes serious plant pests and important biological control agents; they are livestock pests, home pests, and dung recyclers. The most dangerous animals on the face of the planet, as far as humans are concerned, are not lions, great white sharks, or polar bears, but Dipterans—the mosquitoes that transmit malaria and a host of other diseases are responsible for far more human deaths than any vertebrate animal.

*Lepidoptera.* The Lepidoptera, the moths and butterflies (Fig. 10.60), are the fourth mega-diverse order, with about 150,000 identified species. The order name means "scale – wing." Lepidoptera are generally easy to recognize. They have membranous wings clothed in flattened scale-like setae, and most adults have siphoning mouthparts (if they have mouthparts). The origin of the scales is thought to lie in avoidance of spider predation in early moths; the scales stick to the spider's web rather than the wings themselves, allowing the insect to escape. The larvae, called caterpillars (Fig. 10.61), are also easy to recognize, for the most part. They have relatively soft, tubular bodies with well-developed head capsules bearing chewing mouthparts, and usually from two to six pairs of fleshy, unjointed prolegs on the abdomen. The prolegs are typically armed with rows of tiny recurved hooks called crochets that help the caterpillar hold on to the leaf it is eating. Virtually all Lepidoptera feed on plants as larvae, however, a tiny minority are predaceous or parasitic on other insects; virtually all are terrestrial, although a few are aquatic. Most adults, if they feed at all, feed on nectar and other fluids.

**Figure 10.63.** The luna moth.

**Figure 10.64.** A urianid day-flying moth.

Moths are an extremely taxonomically diverse group of organisms. Most can be recognized by their thread-like or feather- shaped antennae; most fold their wings tent-like over their abdomens or hold them straight out to the sides when at rest (Fig. 10.62). The vast majority are nocturnal (though a minority are day-flying). This is an important behavioral adaptation to bird predation. They have other significant adaptions to avoid predation by the night-flying equivalent of birds, bats. Some have tympana ("ears") that allow them to detect bat sonar and take evasive action; some others can actually jam bat sonar with clicks of their own. The long, beautiful tails of the luna moth (Fig. 10.63) aren't just decorations; they apparently confuse bats as well. While most people think of moths as drab and uninteresting, many are actually spectacular insects with beautiful coloration (Fig. 10.64).

**Figure 10.65.** A gray hairstreak butterfly.

The charismatic butterflies are essentially day-flying moths that belong to three superfamilies. Butterflies can usually be distinguished from moths by their more brightly colored wings, propensity to fold their wings vertically over their backs when at rest, and, most reliably, their clubbed or knobbed antennae (Fig. 10.65). The bright colors of many butterflies are important in intraspecific recognition, mating displays, and, in some cases, various defensive strategies. Unlike most other groups of insects, butterflies have a huge number of human fans. Many folks plant butterfly-friendly gardens and butterflywatch much like they may birdwatch. The monarch butterfly, of course, is famous for the spectacular migration we discussed earlier.

Lepidoptera is an extremely important group in terms of ecological end economic impact. Many of out most important agricultural pests are the caterpillars of moth species. Some are medically significant because of the irticating spines and hairs the caterpillars possess. On the other hand, the silkworm has been an important domestic species for thousands of years, many others have been used for food for centuries, and Lepidoptera are important pollinators for many plants.

**Figure 10.66.** A caddisfly. Note the very long antennae.

**Figure 10.67.** A stone-building caddis.

**Figure 10.68.** Twig-building caddis.

*Trichoptera.* The Trichoptera are the caddisflies (Fig. 10.66), and they are very closely related to the Lepidoptera. There are about 12,000 species; the order name means "hair-wing." Caddisfly adults look very much like moths except that their wings are clothed in setae rather than scales, and they typically have thread-like antennae as long as or longer than their bodies. They also have chewing mouthparts if they have mouthparts at all. The larvae (Fig. 10.67) look very much like caterpillars, except they lack prolegs and are aquatic. Caddis larvae are famous for the cases many build by using silk and the substrate in which they live. The cases are characteristic to each species. Some build coiled cases, others straight; some use gravel, while others use twigs, clipped pieces of pine needles, or leaf fragments (Fig. 10.68). Some build no case at all, but spin webs that capture food from the current. Caddis may graze on the algae and other organisms that grow on rocks, prey on other aquatic insects, or feed on plankton strained from the water.

Caddisfly larvae, like the nymphs of the mayflies and stoneflies, generally demand clean, well-oxygenated water, and the diversity of these three orders in a stream can be a very good indication of the general health of the body of water. Caddisflies are critically important prey for trout and other fish.

# Better Communication through Chemistry: Insect Communication

As far as anyone knows, insects don't actually talk like the roaches in *Joe's Apartment* or the ants in *A Bug's Life* and *Antz*, but it is abundantly clear to anyone who has any familiarity with them at all that they do communicate. So, what in the world would an insect need to communicate? What kinds of messages might they need to send? And for that matter, just what is communication?

For our purposes, communication occurs when one organism emits a message of some sort, and another receives it; but here's the important bit: the message has to elicit a response in the receiving individual. If my wife tells me to take the trash out, and I remain, unmoved, in my recliner, watching TV, she hasn't effectively communicated with me (although she most certainly will shortly). Insects, like all animals (and many other organisms), have a host of important things to communicate. They may need to signal the location of a food resource or that a food resource is already fully exploited; they may need to communicate the approach of danger, to rally a counter attack or to flee the danger. Social insects need to communicate the colony's needs so that the division of labor in the colony supports its continued existence; they also need to be able to recognize who belongs and who doesn't so they don't waste the colony's resources on unrelated individuals. Most important for most insects, they need to be able to communicate with potential mates, since an insect that doesn't leave progeny behind (in one sense or another) loses the evolutionary race. Insects may also have very important messages to send and to receive from other species, from plants to vertebrates. Insects use a number of communication modalities, each with advantages and disadvantages. In this chapter, we'll survey these strategies and identify some of the insects that exploit them.

## The Beat Goes On—Sound Communication

Sound communication has a number of advantages, but also several disadvantages. Perhaps the biggest advantage is it is, under the right conditions, omnidirectional—the signal can be propagated in all directions. Exploiting sound also allows for multiple different messages to be broadcast with the same system. Slight differences in sound production methods can also facilitate species discrimination (which we'll address in a moment). However, sound signals dissipate relatively quickly, particularly in complex and crowded landscapes like a forest. The biggest disadvantage, however, is huge—a calling individual is advertising to the world exactly where it is.

While a number of familiar and conspicuous insects (and many unfamiliar and inconspicuous ones) do exploit sound communication, it is important to understand that most insects don't have organs for sound perception and therefore are deaf. However, many of these others can and do perceive vibrations through the substrates they may be standing on. Those insects that do use sound exploit a number of different ways to make their calls.

Probably the most familiar "singing" insects are the Orthoptera, the grasshoppers and crickets. These insects use **stridulation** to make their calls. These insects have a file-like structure across which they draw across a row of pegs, called the scraper; you can emulate this by dragging a pen or pencil (your "peg") across the teeth of a comb (your "file"). The Ensifera, or crickets, katydids, and meadow grasshoppers have their stridulatory structures on the forewings near their base. The file is on one wing and the scraper is on the other (Fig. 11.1). There is often a large, thick plate of cuticle near the file that acts as an amplifier; this structure is the **mirror**. The Ensifera have the tympana they use to hear these sounds on the tibia of their forelegs. The Caelifera or true "short-horned" grasshoppers typically have their stridulatory organs on the hind femur and the tegmina, or forewing (Fig. 11.2). The tympana on these insects are located on the abdomen. In the Orthoptera, typically only the males stridulate, although both sexes can hear the calls. As in many other arenas, the males, with the typically lower investment in reproduction, engage in the relatively risky behavior of singing or calling, while the females respond and choose.

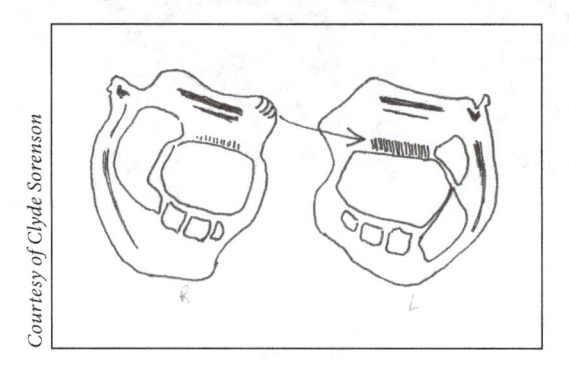

**Figure 11.1.** Schematic of stridulatory organs on forewings of a field cricket.

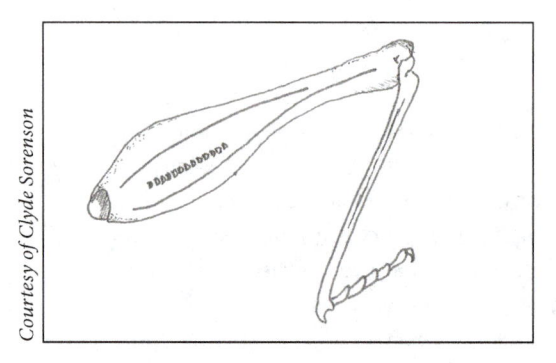

**Figure 11.2.** The stridulatory pegs on the leg of grasshopper.

**Figure 11.3.** The tip of this predaceous wheel bug's rostrum (beak) is resting in its stridulatory groove.

Even though these sound-making organs are rather simple, they can be used to elegantly send many messages, by varying the rate, duration, and pattern of the stridulatory movements. Common field crickets have one call to advertise their location, another to court any female attracted, a third to signal consummation, and a fourth to warn competing males that approach too closely. Similarly, small variations in the structure of the stridulatory organs, coupled with variations in how they use them, allow very closely related species to have their own unique sets of calls.

Orthopterans aren't the only insects that stridulate. Many Hemipterans, such as some assassin bugs, stridulate by dragging the tip of the rostrum (the tip of their beak-like mouthparts) against a file on the underside of the thorax between the front legs (Fig. 11.3). These sounds may be important in intra-species communication, but can also serve as warning sounds to potential predators. Many ants and some other Hymenoptera stridulate; in the ants and velvet ants (actually large, wingless wasps), a scraper on the rear portion of the abdomen is drug across a file on the front part of the abdomen (Fig. 11.4). Many beetles also stridulate, often by scraping the wings against the abdomen; even some beetle grubs can stridulate.

100   Insects and People

Another impressive sound-making organ is the **tymbal** of male cicadas (Fig. 11.5). Tymbals are ridged plates of cuticle located on the upper, front side of the abdomen (Fig. 11.6); they often have a protective plate partially occluding them. The plates are attached by a sturdy strut of cuticle to massive, often asynchronous, muscles located in the abdomen, and much of the abdomen is occupied by air spaces that serve to amplify the sounds produced by the tymbals. To create sound, the large muscles deflect, or bend the tymbal. When the plate (and each of the ridges in it) bends, it produces a pulse of sound, much like when you take the lid from a pickle jar and deflect it with your fingers to make it "pop." Just like the lid, the tymbal and its ridges make another sound when they pop back to their original shape upon relaxation of the muscle. With this relatively simple mechanism, cicadas are able to make some incredible sounds. The loudest insect in the world, in real terms, is a cicada that can produce sounds in excess of 105 decibels, levels approaching the racket that a table saw or a heavy metal concert (roughly equivalent noises in terms of annoyance) might produce. Some of the huge cicadas from tropical Southeast Asia are undoubtedly louder. They are also able to produce some otherworldly sounds ranging from a decent imitation of a fire truck or ambulance, to the pulsating whine of a science-fiction movie space ship. Cicadas can use their tymbals defensively; many have very loud alarm calls that can frighten birds and rodent predators away.

Some moths also have tymbals, but these are located in the thorax. Moth tymbals make clicking sounds that help them avoid bat predation, by either jamming the bat's sonar or by informing the bat of the moths' toxicity.

A number of insects make percussive sounds by banging a part of their body against the substrate they are standing on, although many of these signals are too faint to be perceived by human ears. Termite soldiers of many species use head rapping to signal alarm should the colony come under assault. Green lacewings and many stoneflies make drumming courtship calls with their abdomens. Treehoppers make similar drumming sounds with various parts of their bodies. In most of these cases, the receiving insects feel these vibratory signals through their feet rather than hearing the airborne sounds. A group of tropical butterflies, the crackers (Fig. 11.7), makes percussive sounds

**Figure 11.4** Hymenopteran stridulation.

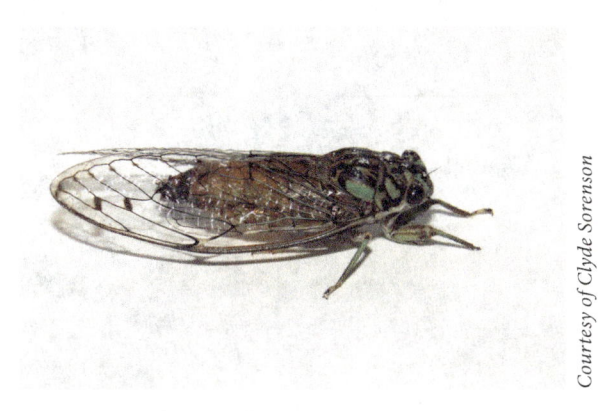

**Figure 11.5** Male hieroglyphic cicada.

**Figure 11.6.** Tymbal (ridged structure under wing) on hieroglyphic cicada.

by whacking their wings forcefully together, and these are used primarily to startle and confuse potential predators. The males of some of these species also make clicking sounds that are thought to be courtship calls.

A few insects make hissing or whistling sounds by forcefully expelling air from their spiracles; the Madagascan hissing cockroach is probably most noted for this behavior. The death's-head hawkmoth (Fig. 11.8) is unusual in that it actually makes a crackling, hissing sound by forcing air through its mouthparts. This moth regularly steals honey from honey bees, and at one time it was thought these sounds were to calm the bees.

Mosquitoes, midges, and a number of other insects use the humming sounds their wings produce as part of their courtship. Male and female mosquitoes must actually harmonize their wing beat frequencies in order to proceed with mating. Some parasitoidal wasps make quite complicated songs by vibrating their wings while approaching a potential mate on foot. The famous waggle dance of the honey bee is another example of a vibrational communication mode accomplished with the wings. Honey bee foragers signal direction of a potential nectar source by the orientation on the comb face in the hive of the "waggle" portion of the figure-eight dance (Fig. 11.9). They signal distance by the duration, or length, of this portion of the dance. It's important to remember that the other perceive this rather elegant message by feel, as the vibrations of the wings during the waggle are transferred to the air and the wax of the comb. The other bees don't watch this dance because it usually occurs in the virtually complete darkness of the hive interior.

**Figure 11.7** A cracker butterfly.

**Figure 11.8.** Death's-head hawkmoth.

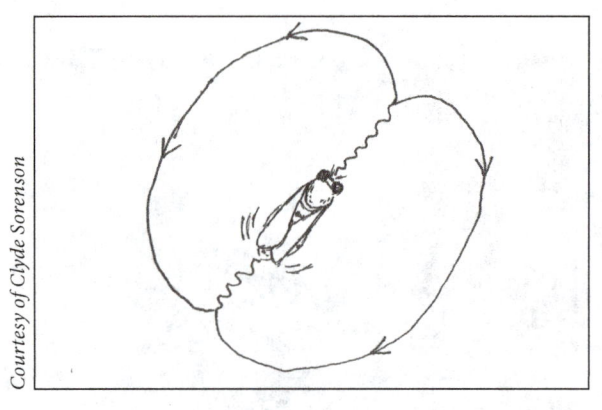

**Figure 11.9** Diagram of the waggle dance of the honey bee.

# The Eyes Have It— Visual Communication

A great many diurnal insects use patterns of markings and color on their wings to communicate their species identity and sex. Many, like some of the fruit flies (Fig. 11.10), couple patterned wings with stereotypical movements as an important part of their courtship routine. Butterflies, in addition to the colors we can see, often have ultraviolet reflective patterns (Fig. 11.11) on their wings that twinkle in characteristic patterns (to those, like other butterflies, that can see it). Other insects, including flies like midges and some mayflies, mass in aerial **leks** to provide a visual cue to females seeking mates (Fig. 11.12).

**Figure 11.10.** Apple maggot fly, demonstrating patterned wings.

Perhaps the most dramatic and fascinating visual communication systems in insects involve **bioluminescence**. Bioluminescence is the production of light by living organisms, and a surprising number of organisms are able to do this. While some insects that exploit light hijack light-producing microorganisms, most produce their own through a really nifty chemical reaction. The primary chemical involved in the reaction is luciferin. Luciferin reacts with an enzyme called luciferase, in the presence of oxygen and magnesium. The resulting degradation product of the luciferin emits light. The insect can turn the light on and off by controlling the flow of oxygen to the reaction through the abdominal trachea. After the reaction, the degradation products can be recycled to make more luciferin to resupply the reaction. There are actually a large number of variations of luciferin and luciferase, and these can produce different colors of lights in different internal environments in different insects. This light production system is exceptional because almost all the energy that goes into it as chemical energy comes out as light, with no energy lost as waste heat.

The most famous bioluminescent insects are the fireflies (or lightning bugs), which are actually soft-winged beetles. Fireflies are found in many humid parts of the world; several species are common in eastern North America. Fireflies use light to find mates; each species has its own distinct signaling system to ensure they mate with members of their own species. In most the males fly about at the right time of the night (often different for different species), flashing the appropriate signal; in one common eastern species, *Photinus pyralis* (fig. 11.13), the males make a "J" shaped, swooping

**Figure 11.11.** a. An orange-barred sulfur photographed under visible light; b. the same insect illuminated by u.v. light. Note that the orange patches on the forewings and edges of the hindwings strongly reflect u.v., and that the white patches under them also intensely reflect u.v.

**Figure 11.12.** A swarm of midges leking over a statue in a cemetery.

**Figure 11.13.** *Photinus* firefly.

**Figure 11.14.** *Photurus* female firefly eating a male *Photinus* firefly.

yellow-green flash every six seconds as they fly about just after dusk, two to four feet above the ground. Females sit in low vegetation and wait to spy a male; when one sees him, she waits two seconds, and flashes back. The male responds to this while flying closer, and, after several iterations, the male lands and runs over to the female, where she makes her final assessment of his quality, and they mate. (Males typically provide a nuptial gift.)

Another group of fireflies exploits light communication in a different way. In these species, in the genus *Photurus*, virgin females do as they should, and watch for, and signal to, males of their species. Once they mate, however, a female *Photurus* watches for a cruising male *Photinus*. When she sees one, she mimics the flash of a female *Photinus*, calls the male down, and then eats him (Fig. 11.14). She gets an important spider-repelling chemical from the male *Photinus* dinner; the more she eats, the safer she and her eggs are from spider predation.

In some parts of the world, notably parts of Southeast Asia and the southern Appalachian mountains, large aggregations of male fireflies flash in unison, producing a spectacular display. One line of thought suggests that they synchronize to avoid losing track of replying females in all the light confusion.

Other insects also exploit bioluminescence. Several species in various parts of the world are called "glowworms," although they may be very different kinds of insects. North American glowworms are the larviform females of some very odd beetles in the family Phengodidae (Fig. 11.15). They have rows of green light-producing organs along the sides and across the backs of their long bodies, and they glow continuously through the night as they crawl around the forest floor. They look like three-inch long subway cars easing through the woods, advertising how bad they might taste to potential predators.

The glowworms of New Zealand and Australia are actually the predaceous larvae of a fungus gnat. They live on the ceilings of caves and grottos, where they construct silken snares consisting of numerous silken lines clothed in sticky mucus. They have a blue light-producing organ at the tip of their abdomen; hundreds together on the ceiling of a cave look like a beautiful, starry sky. The lights attract other insects, which fly up into their sticky lines. The larvae then haul them up and eat them. This may not actually be interspecific communication in the purest sense, but it is still fascinating!

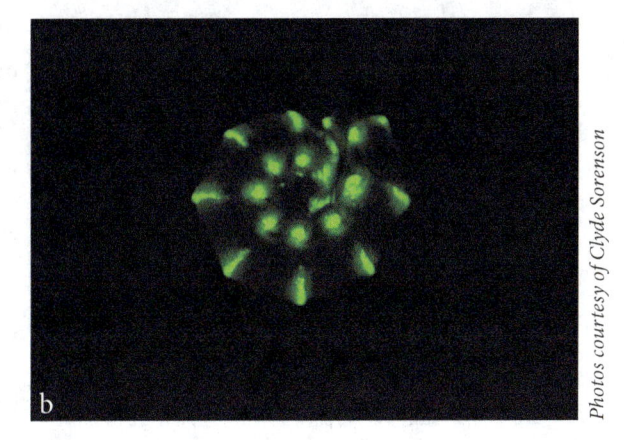

**Figure 11.15.** a. female Phengodid beetle; b. showing light organs.

# I Smell Sex and Candy—Chemical Communication

Far and away the most common form of communication across insects is chemical communication—using scent to send signals. The advantages to chemical communication are significant; chemical messages can be very specific, helping to ensure that only those individuals an insect desires to communicate with respond to the message. Unlike sight- and sound-based signals, predators may have difficulty in detecting chemical signals. Chemical messages can also, on occasion, travel relatively vast distances and still elicit a response. Some male moths appear to be able to detect females of their species from miles away. On the other side, long-range chemical messages are unidirectional—they can only travel downwind. Beyond wind direction, weather can have a huge impact on whether or not a chemical message can be sent or whether it will be received.

There are several different functional categories for chemicals used to communicate, in a very broad sense; some of these may facilitate long-range communication, while others may require contact between the two communicators to transmit the message. Chemicals that are used in communication are called **semiochemicals**. Semiochemicals are further divided by whether they facilitate interspecific communication (between different species) or intraspecific communication (between members of the same species), and also by the nature of the message. Interspecific messages that are positive for the sender and negative for the receiver are called **allomones**. The noxious spray of a ticked-off skunk is the classic example of an allomone. Once that animal unleashes its sulfurous weapon, it calmly walks off to do what it was previously set on doing while the recipient is left writhing on the ground in pain, typically with no further intention to harm the skunk. A great many insects produce allomonic chemicals for defensive purposes, and a great many plants produce allomones to protect themselves from insects and other herbivores.

A chemical message that is negative for the sender and positive for the receiver is called a **kairomone**. A good example of a kairomone is the carbon dioxide we exhale with every breath. We can't help but produce this chemical in order to sustain life, but some parasitic arthropods like ticks and mosquitoes use $CO_2$ (and other odors) to locate us. Kairomones are again abundant in the natural world. An interesting thing to acknowledge, and to ponder, is that a single chemical produced by a living thing can function as both an allomone and a kairomone to different species. For example, the chemicals that give cucumber its bitter taste are produced as allomones by the cucumber plant to protect it against most herbivores. These chemicals are distasteful, and sometimes toxic, to most

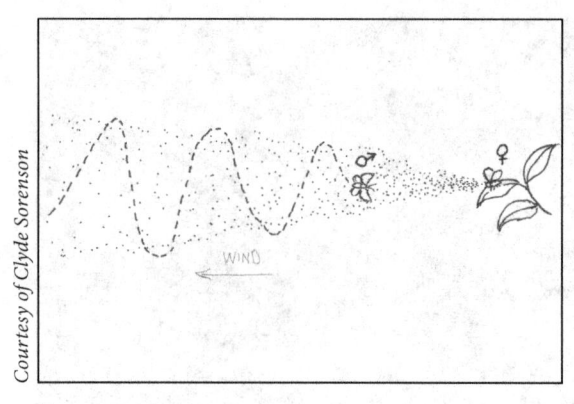

**Figure 11.16.** Schematic of a pheromone plume moving downwind from source female.

plant eaters. However, for those herbivores that have evolved to exploit cucumber plants, including several insects, these same chemicals are how they identify the plant as a suitable food, and so the chemicals are kairomones with respect to these specialist herbivores.

A chemical message that is positive for both the sender and the receiver is called a **synomone**. The fragrance a night-blooming flower produces to attract pollinating moths is a great example of a synomone. The flower gets pollinated, and the moth gets a sip of nectar—both win. Synomonic chemicals that relay messages within species are called **pheromones**, and almost all insects exploit pheromonic communication to some degree.

Pheromones are used to send a wide range of messages. Probably the most widely used pheromonic signal is for mating. These are called **sex pheromones**. In many insects, the female produces a long-range sex pheromone, often from a relatively safe, secluded spot, while the males engage in the risky behavior of coursing through the air in search of the message. Once a male encounters an appropriate pheromone plume, it turns up wind, and follows the intensifying signal, bouncing off the edges of the plume as he follows a zigzag course to the source (Fig. 11.16). Additional pheromones may be involved once the male finally finds the female, and, of course, there may be exchanges of nuptial gifts.

In some, like the boll weevil, the males produce a pheromone that attracts both females and males, and this functions as an **aggregation pheromone**. In the case of the boll weevil, a male can't produce the pheromone until he feeds on fresh cotton squares (flower buds). This signals to females that the male has found a food source and a substrate for egg laying; other males respond to it because there may now be females in the area. Bark beetles also produce aggregation pheromones. In this case, large numbers of beetles must attack a tree in unison to overcome the trees defenses. Bark beetle pheromone systems are more complicated. Once enough beetles have been recruited to the tree, they begin producing another signal that repels any additional recruits since the tree can only support so many. (Interestingly, the bark beetles use chemicals they acquire from the host tree to produce both the aggregation and repelling pheromones.)

Female insects may also produce **epideictic pheromones**, which inform other females that a resource has already been exploited. The female blueberry maggot fly drags her ovipositor over the surface of a fruit in which she's just laid an egg, depositing a signal to other females that there is no point in wasting an egg on this already occupied berry.

A great many insects exploit alarm pheromones to warn conspecifics of impending danger. In most social Hymenoptera, including honey bees, ants, and many wasps, these signals usually rally other colony members to attack the invader, whatever it might be. Anyone who has accidentally stepped into a red imported fire ant mound has witnessed an extremely effective alarm pheromone, as the ants come boiling out of the soil. In this instance, the signal propagates throughout the colony extremely rapidly because alarmed individuals also release alarm pheromone. In insects that aren't armed with effective defensive weapons like stingers, an alarm pheromone causes the receiving individuals to flee or retreat. In many aphids, an alarm pheromone signal causes the insects to drop from the plant to the ground, where the predator that elicited the alarm is less likely to find them.

Social insects also exploit pheromones to manipulate the composition of the colony. In a termite colony, workers are constantly receiving messages from the other workers, soldiers, and immatures they encounter, and they relay these messages to the queen. She subsequently produces pheromonal messages that direct the colony. If soldiers have declined because of losses from defending the colony, the workers who tend the queen will relay this information chemically to the queen, and she will send a message to the colony to raise more soldiers. Termites and most ants also make extensive use of trail-marking pheromones, which steer colony mates to food resources scouts have identified.

If the queen in a honey bee colony is getting old, her pheromone production declines, and this is the signal to the workers of the hive to produce a batch of new queens to replace her. The workers eventually kill the old queen, and once they begin emerging from their pupal cells, the new queens fight until only one is left. The survivor leaves the hive to mate. Once fully mated, she returns and settles into the egg-laying role her mother once had.

## The Enemy Within— Using Pheromones to Control Pest Insects

Insects have been the model organisms in our understanding of pheromones, but we've only been studying them intensively for about 50 years. We've learned that, in many insects, pheromones aren't just single volatile chemicals, but rather very specific blends of several chemicals. We have identified sex pheromones for almost 2,000 species of insects, and aggregation and other pheromones for quite a few others. Many of these insects are important agricultural pests, and knowing the pheromone blend for such an insect can be a powerful tool. There are several ways we can exploit synthetic versions of pheromones to help reduce the impact of pest insects.

Perhaps the most common and widely deployed tactic is to use pheromone-baited traps to **monitor pest populations**. Different kinds of insects require different kinds of traps. Moths that tend to fly up to escape, for instance, are best captured in an inverted cone trap (Fig. 11.17), while those that fly downward to escape can be trapped in a funnel-topped bucket trap (Fig. 11.18). Some

**Figure 11.17.**

**Figure 11.18** Bucket-type moth pheromone trap.

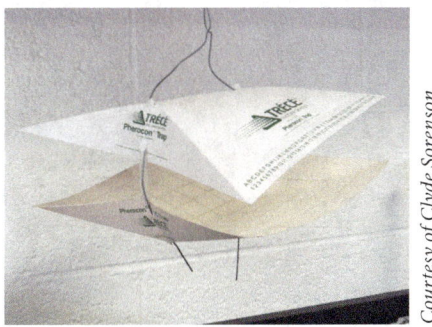

**Figure 11.19.** Wing-type pheromone trap.

Better Communication through Chemistry: Insect Communication    107

**Figure 11.20.** Carton-type gypsy moth trap.

**Figure 11.21.** Bark beetle trap.

**Figure 11.22.** Pheromone "puffer" for dispersing codling moth pheromone in apple orchards.

small moths and beetles can be effectively captured in cardboard wing traps, coated on the inside with a sticky adhesive (Fig. 11.19). Gypsy moth males are captured in specially designed cardboard traps (Fig. 11.20). Bark beetles respond best to traps constructed of black funnels or panels vaguely reminiscent of a tree trunk (Fig. 11.21). In all these cases, trapping doesn't effect population control, but rather provides us information we can act on.

Populations of some insects, however, can be effectively reduced with pheromones through an approach called **mass trapping**. This works best for those insects that have pheromones that elicit responses from both sexes, like the boll weevil. In this case, a very high density of traps can substantially reduce the local population. Mass trapping can also be used to good effect in interiorscapes like greenhouses.

A variation on this approach is the **attract-and-kill** strategy. Pheromone dispensers can be combined with a toxicant so that attracted insects receive a dose of insecticide and subsequently die. This strategy reduces the amount of insecticide that must be distributed in the environment and specifically targets only the pest. This tactic can also be used with a food attractant bait, which is a common practice for targeting some household pests like cockroaches.

A third strategy is **mating disruption**. With this tactic, we attempt to flood the environment with pheromone by distributing it throughout the area we're trying to protect with multiple dispensers (Fig. 11.22). A male moth in such an environment will have great difficulty finding calling females, because clearly defined pheromone plumes can't form if pheromone is everywhere. If the males can't find the females and mate, the females can't lay fertile eggs that produce the damaging caterpillars. This has been used to very good effect in apple orchards for important pests like the codling moth and oriental fruit moth. It's also been successfully deployed in other crops.

Pheromones and other semiochemicals are likely to grow in importance in pest management in the future, as we learn more about them and develop additional tools for their deployment.

# XII

# Big Queens and Little Kings: The Social Lives of Insects

The vast majority of the species of insects in the world aren't much interested in others of their kind unless there is a potential for reproduction. But, as you might expect, and probably already realize, some are highly social. There is a continuum of sociality in insects from solitary to eusocial ("truly social"). It's important to remember, however, that the categories that follow are somewhat arbitrary. Some insect species exhibit behavior that is intermediate, in some respect, to each of the pairs of categories, and some species may fall into one category in some places and times and a different category in others. In other words, insect sociality is very complex, and our understanding of it is constantly being modified. All insects that demonstrate some level of sociality below eusociality are considered **parasocial**.

**Solitary** insects typically only associate with each other to mate, or to exploit a concentration of food or another resource. Perhaps the epitome of the solitary insect is the preying mantis (Fig. 12.1). Unless they are mating, mantises typically regard each other only as food. Huge aggregations of "solitary" insects can occur. A mayfly swarm may have thousands of males in it, but each is pretty much out for himself. In most solitary insects, the females provide no parental care once they lay their eggs or give birth.

**Communal** insects share a common nesting area and may associate with each other when foraging, but each reproduces on its own. One good example of a communal insect is the tent caterpillar (Fig. 12.2). The group of caterpillars emerging from an egg mass makes and shares a common silken nest, but the individuals disperse to pupate. Some of the soil-dwelling andrenid bees are also communal. Large numbers of bees aggregate in certain soil banks to build their nests, but each female makes her own tunnel and takes care of provisioning her own nest and larvae.

**Subsocial** insects provide some level of parental care to offspring. Quite a few insects are subsocial, and the parental care may be provided by the female alone or by a mated

**Figure 12.1.** A Carolina mantid, an excellent example of a solitary insect.

Big Queens and Little Kings: The Social Lives of Insects   109

**Figure 12.2.** A communal nest of Eastern tent caterpillars.

**Figure 12.3.** *Cryptocercus* cockroaches, examples of sub-social insects.

**Figure 12.4.** A female treehopper and her offspring.

pair. A classic example is the *Cryptocercus* woodroach (Fig. 12.3). These close relatives of termites feed on wood and excavate galleries in dead logs. A mated pair cares for and feeds a small batch of nymphs until they are almost grown. The almost grown nymphs eventually go off on their own to pair and produce their own young. Other well-known subsocial insects are the dung beetles, carrion beetles, many treehoppers (Fig. 12.4), and lace bugs.

**Quasisocial** insects live together in a common nest and provide communal care of offspring produced by all the colony members. This is a relatively rare form of social organization found in some bees.

**Semisocial** insects live in the same nest and cooperatively care for the immatures, but only one of the colony members is fertile and produces all the offspring. This is the first instance where a division of reproductive labor occurs. Some sweat bees and a few wasps (Fig. 12.5) exhibit semisociality. Most of these species have an annual cycle, and some can switch to other forms of sociality.

**Eusocial** insects live together in the same colony, collectively care for the immatures, have a reproductive division of labor (and some times other castes), and have overlapping generations. Many of these colonies are perennial, and some can persist for decades. All termites, all ants, and many bees and wasps are eusocial, and there is evidence that some other insects are eusocial as well. While the proportion of total insect species that are eusocial is small, these insects have tremendous ecological impact—almost certainly more than any other suite of insects. Eusocial termites, ants, and bees share some important characteristics but also differ in important ways.

All eusocial insects exhibit **polymorphism**, which means that the colony includes members that look different from each other, sometimes radically so. In all, the reproductive female, the queen, is typically significantly larger than the workers, because, at the very least, she must have a large abdomen to accommodate the huge ovaries need to supply the colony with offspring. Male reproductives tend to be larger, as well. In some, particularly some ants and termites, there's even more polymorphism, because these colonies have recognizable soldier castes as well as workers (Fig. 12.6). There are often major (big) and minor (little) workers, and sometimes major and minor soldiers. In most species with soldier castes, the soldiers have mandibles so modified for defense that they must be fed by workers.

In most eusocial colonies, the queen dominates the colony, and the messages she receives and sends dictates the colony's fate. The queen is, in many ways, the most important member of the colony because she produces the replacements for colony members. In some species of termites and ants, there can be more than one queen (sometimes many more) in the colony. Most colonies have mechanisms for replacing queens, but the catastrophic loss of the queen can ultimately mean the death of the colony in many species. It is important to realize that the queen doesn't actually command any individual's activities. Her influence on the colony is directed by the colony as a whole responding to her chemical messages.

Almost all eusocial insects engage in **trophallaxis**, which is the sharing of gut contents. This behavior is critically important in termites because this is the way the microorganisms that help digest the cellulose in wood are transferred from one insect to another. Termites exchange gut contents when one individual feeds on material from another's anus. This is known as **proctodeal** trophallaxis. Ants and bees exchange gut contents mouth to mouth. This is called **stomatodeal** trophallaxis. While ants and bees engage in this behavior to exchange nutrients, it is also an important way for members of colonies of all eusocial insects to chemically exchange information.

Members of eusocial colonies of insects spend a great deal of time **grooming** themselves and their colony mates. While hygiene is critical for animals living is such high densities and in such constant contact, grooming is yet another opportunity for colony mates to exchange information.

Colonies of social insects typically develop colony-specific **nest odors** that allow individuals to identify colony-mates from strangers. Wasting resources on unrelated individuals could harm the colony, so strangers are typically killed or at least harassed until they leave. However, some ants form supercolonies that can span miles, because of genetic similarity, and in these supercolonies, individual ants can freely move from one colony site to another.

**Figure 12.5.** *Polistes* paper wasps; these creatures exhibit semisociality.

**Figure 12.6.** Termites, demonstrating polymorphism. All these individuals came from the same colony; the picture includes workers, soldiers, and primary and secondary reproductive.

Most eusocial species construct **complex nests** and are able to maintain a very stable environment in them. Honey bees maintain the brood nest, where larvae and pupae reside, at about 95° F. Special workers called "heater bees" shiver wing muscles to produce heat if it's needed. If the temperature is too hot, they use their wings to fan fresh, cooler air into the hive and to evaporate water spread on the comb. Many soil-dwelling ants erect cone-shaped mounds that they can use to capture heat (Fig. 12.7). Termites, too, precisely control the temperature inside their nests by regulating airflow. Many ants and termites also exploit the heat generated by decaying vegetation to help warm their nests.

**Figure 12.7.** Wood ant (*Formica*) mound in a Swedish forest.

Termites, ants, and honey bees all use mating flights to establish new colonies. Within a species, the new reproductives (future queens and drones) leave the colony to mate. In honey bees, the newly mated queen returns to the hive. Hive number is increased by swarming, where the old queen leaves with some proportion of the hive's workers to a new location. In ants and termites, mated queens (accompanied by male "kings" in termites) leave to establish new colonies.

There are also some major differences between these three groups of eusocial insects. Termites, of course, feed on wood and other vegetation. Honey bees eat nectar for carbohydrate energy and pollen for their protein needs. Ants are extremely diverse in what they eat; some species are completely carnivorous, while others feed exclusively on certain fungi, and still others are quite omnivorous.

Termites, unlike honey bees and ants, have conventional sex determination, and this has a substantial effect on termite colonies compared to those of the Hymenopterans. In most species the male reproductives remain with the queen(s) and apparently re-mate with them throughout their lives. In ants and honey bees, the male reproductives have no use after the queen is fully mated and aren't tolerated in the nest after mating. In termite colonies, workers and soldiers are of both sexes, while, of course all the workers in the Hymenoptera are sterile females. Since termites are hemimetabolous, with nymphs that look like miniature versions of the adults, nymphs can and do participate in the maintenance of the colony, while the larvae of the holometabolous ants and honey bees are legless creatures that must be tended by the workers.

# XIII

# Good Guys and Goodies: Insect Products and Services

Insects abound. They dominate in numbers, diversity, and often in biomass. It's not surprising, then, that humans have figured out ways to utilize some of them. As it turns out, we get quite a few products from insects and made of insects, and, perhaps more importantly, they provide innumerable valuable services. In this chapter, we'll discuss some of these goods and services and their history.

Before we delve into more specific cases, it's important to clarify that virtually all insects are beneficial. Of the 1,000,000 or so species that have been described so far (and, remember, there are millions yet to describe), only a few hundred to a couple thousand are chronic pests for someone somewhere. The rest are extremely valuable to us, if for no other reason that they help maintain the webs of life that we rely on for our survival. The ecological services provided by insects are immeasurable. E. O. Wilson, one of the greatest biologists of this time and perhaps any time (and a specialist in ants) said, "If all mankind were to disappear, the world would regenerate back to the rich state of equilibrium that existed ten thousand years ago. If insects were to vanish, the environment would collapse into chaos." In spite of our incredibly obvious presence on this planet, and in spite of all the things we have done to reshape it, we are, in the end just one run-of-the-mill species of primate. But insects are the threads that hold life's creation together, at least on land and in freshwater. Now, enough with the pontificating. Let's talk about some insect products.

## The Many Gifts of the Honey Bee

Certainly the most well-known insect product is honey. Many other bees produce honey of one sort or another, but virtually all the honey humans consume is produced by one species, *Apis mellifera*, the Western honey bee (Fig. 13.1). This species is native to Europe, Africa, and western Asia, and there are numerous subspecies. Most domesticated honeybees are descendants of European subspecies. Honey bees make honey by harvesting nectar from flowering plants, mixing it with enzymes in their crop, which changes the sugars in the

**Figure 13.1.** Western honey bee, *Apis mellifera*. The queen is the large individual near the middle of the picture with the orange, largely unmarked, abdomen.

**Figure 13.2.** Beekeeper working with moveable frame bee hives.

| Table 13.1 | A Few Industrial Uses of Beeswax |
|---|---|
| Dying Processes | Cosmetics |
| Lost-wax jewelry casting | Resistance painting |
| Lubricant | Dental wax |
| Shoe polish | Pattern-making |
| Candles | Lithographic ink |
| Culinary uses (candies, etc.) | Modeling |
| Fly-tying | Floor wax |
| Tailoring | Varnish |

**Figure 13.3.** A feral bee colony which has built comb of wax on a tree limb.

**Figure 13.4.** The worker bee near the center of this picture has pollen baskets (corbiculae) full of pollen.

nectar, and dehydrating it in the bee comb. Once it has reached the appropriate moisture content, they cap the cells with wax. Humans have kept bees for perhaps four thousand years, but, for much of that time, harvesting honey meant destroying the hive. In the 1800s we developed movable frame technology (Fig. 13.2) that allowed beekeepers to take honey while preserving the bees. For most of our history, honey has been the only readily available sweetener for most people. The ability to economically refine sugar from sugarcane and sugar beets is a relatively recent development; up until about 500 years ago, sugar was too expensive for common folk. Honey is an excellent sweetener. It can be stored for long periods of time (virtually forever), has complex sugars, can be reliquified if it crystalizes, and actually kills microorganisms that might try to colonize it (by desiccating them). (This desiccating ability has also been exploited medicinally in dressing small wounds.) Honey is still the preferred sweetener in many applications, and many people greatly appreciate the complexity and variety of flavors that can be found in different kinds of honey.

Honey bees also, of course, produce wax (Fig. 13.3). Beeswax is produced by glands on the abdomen of honey bees. They groom the wax scales from their abdomen and then form it into pellets, which can be shaped into the comb. In the wild, the bees construct the entire structure of their colony home from wax, usually inside a tree cavity or other hollow. Beeswax is composed of a complex of esters and other compounds and, while it is obviously useful to the bees, we put it to a tremendous number of uses as well. Beeswax is edible, flammable, and an excellent lubricant. Table 13.1 identifies some of the many ways we use this remarkable substance.

Honey and wax are not the only commercial products we acquire from bees. Bees collect pollen to provide the protein requirements of the colony since honey is almost pure carbohydrate in the form of sugars. Bees are "hairy" in part to facilitate pollen collection. They groom the pollen that collects

114   Insects and People

in the setae on their bodies and pack the sticky particles into basket-like structures on their hind legs called corbicula (Fig. 13.4). We collect the pollen pellets by placing over the hive entrance a vertical frame of wires spaced far enough apart to allow the bees to pass through, but not the bees and the pollen pellets. The pellets pop off into a collection tray located below. Pollen is used by many as a nutrient supplement; however, while it is protein-rich, it can be a potent allergen to some people.

**Propolis** is another bee product. It is composed of sticky plant resins the bees collect to make a "glue." They use it to seal cracks in the nest cavity they occupy. Propolis, among other uses, has been incorporated into varnishes. It was a critical component of the varnish Stradivarius used on his remarkable stringed instruments. It also may have some important medical applications, and a great deal of research is being conducted on its medicinal value. **Royal jelly** is a protein-rich substance the bees feed to larvae—queens receive much more as they develop. Royal jelly is purported to have a great many health benefits, although few have been documented by controlled research. Even bee venom has some therapeutic value as a treatment for arthritis, and it may have some applications in the treatment of cancer.

# The Smallest Domestic Livestock: Silkworms and Silk

Many insects produce silk, but caterpillars, the larvae of moths, are most known for the ability. Caterpillars use silk to build protective cases called cocoons, within which they pupate. The cocoon is not composed purely of **fibroin** fiberous silk. The strands are glued together with a protenaceous substance called **sericin**. Emerging moths secrete a substance that dissolves the cocoon at one end, allowing the insect to escape.

The creature that produces the silk of commerce, *Bombyx mori* (Fig. 13.5), is a completely domesticated insect. It does not occur in the wild and it cannot survive without the assistance of humans, much like the leghorn chickens that produce the eggs and meat we eat, and the Holstein cows that produce the milk we drink. The silkworm was first domesticated in China over 5,000 years ago. Domesticated silkworms have massive silk glands compared to their wild relatives, and the adults are incapable of flight.

**Figure 13.5.** Mating silkworm moths.

Silk is an incredible substance, given that it is basically hardened caterpillar spit. It is composed of a protein called fibroin arranged as paired filaments. Fabric made from silk is very light and extremely smooth, and, because the fibers themselves are triangular in cross-section, the fabric has a famous sheen and sparkle. It holds heat and feels wonderful against the skin. It retains dyes very well and can be used to produce incredibly vivid yarns. It is also one of the strongest fibers found in nature, and was once incorporated into the armor of Japanese warriors.

The commercial production of silk is called sericulture. Since silkworm caterpillars only eat the leaves of the white mulberry tree, silk production begins with the cultivation of mulberry plants. It takes about 200 pounds of mulberry leaves, fed to about 2,500 caterpillars, to produce one pound of finished silk yarn. Hatchling caterpillars are fed on tender, young mulberry leaves placed on screens to allow the frass to fall through. As they grow, they are transferred to new beds of leaves by carefully rolling the screens and then spreading them out on the new foliage. As in most leaf-feeding caterpillars, the last larval instar eats about 80% of the total foliage consumed, so as they grow, it becomes quite a task to keep them supplied with fresh leaves.

**Figure 13.6.** Silkworm cocoons in cardboard spinning frames.

As the caterpillars reach larval maturity, they are transferred to open-sided cardboard cartons that provide them a place to spin (Fig. 13.6), and, over the course of about 24 hours, they spin their cocoons. Once completed, the cocoons are punched out of the cartons, and then placed in boiling water to soften the sericin and allow the double-stranded, single filament to be teased off and unspooled. Since an individual filament is too fragile to make thread, the filaments from about 10 or so cocoons are combined and unspooled together. The filament from each cocoon is about one third of a mile long. Several of these combined threads are then spun to make the fine-diameter yarn used to make silk fabrics. The killed pupae don't go to waste. In most silk-producing countries, they are used as food.

Silks are, of course, produced by many other arthropods, and a great deal of research attention has been devoted to figuring out ways to exploit these other silks. Spider silks are even stronger than silkworm silk, and could have some remarkable applications, but, of course, spiders don't produce the same volume of silk and they are much more difficult to rear and to get silk from. Some are working on ways to produce spider silks in silkworms through genetic engineering.

## Shellac, Red Dye, and Blue Ink: Other Insect Products

While honeybees and their products and silk are well-known, we have derived a number of other useful products from insects. One of these is **shellac**. Shellac is a hard, waxy, substance produced by soft-bodied, sessile scale insects found on certain trees and shrubs in the Indian subcontinent. The insects, called lac insects (Fig. 13.7), produce the substance to protect them from predators and the harsh environment they live in. We harvest it by scraping the insects and their covers from the twigs and branches of the shrubs. Shellac is refined from this mixture of insect bodies and scale coverings and then processed into transparent flakes that can then be used for a range of purposes. Like beeswax, shellac is essentially a naturally produced plastic. Since it is a polymer, shellac can be used to create an excellent, high gloss furniture finish. It is also used for culinary purposes (it is edible), to give a high sheen to candies, other foodstuffs, and even pills. It has a large number of other industrial applications where a smooth, waterproof coating is required.

**Figure 13.7.** Raw shellac from lac scale insects.

A native Central and South American scale is the source for carmine red dye, or **cochineal**. This scale lives on prickly pear cacti, and it produces a substance called carminic acid to protect it from its natural enemies. Carminic acid is brilliantly red in color, and, when properly formulated, is an excellent textile dye. Cochineal produces a very stable and permanent red. Native Americans exploited this dye for many hundreds of years before the European exploration of the Americas. Europeans quickly adopted cochineal, since it produced a more vivid and intense red than virtually any dye they previously had. The product that reached Europe was the dried bodies of the scales. The hard, graphite-colored little particles resembled nothing living and were long regarded as a mineral product. In the early days of the European cochineal trade, the material was worth more than silver. Cochineal was used to die the brilliant scarlet coats of the officers in the British Army (the Red Coats of Revolutionary War fame), and it has also been used as an artist's pigment. Today, cochineal is widely used in the cosmetic and food industries. If you use a red lipstick, it is quite likely to contain this insect-based product.

For centuries, the gold standard for permanent ink was another insect-based product, **iron-gall ink**. Iron-gall ink is made by mixing a preparation of tannic acid with a solution containing iron. The source of the tannic acid is galls produced by tiny wasps on oak leaves (Fig. 13.8). The wasps lay their eggs in the leaves early in the spring, just as they are beginning to expand, and the round, hard galls form in response to chemicals produced by the wasp larvae developing inside. The gall provides both nutrition and protection for the larva. The tree, in response to the invasion of its leaf tissue, deposits large amounts of tannic acid into the gall in an attempt to kill the invader. This ink has long been preferred for legal documents since it is indelible, and it still has significant use for this purpose, although it may be formulated from other acid sources. If improperly made, the residual acid in the ink can damage paper, but if made properly it can last quite literally for centuries. Most of the surviving canon of Medieval European literature was inscribed, by hand, with iron-gall ink.

**Figure 13.8.** Oak leaf galls.

## Insects Feed the World

In addition to supplying direct products to us, insects are valuable for a great many ecologically and economically significant services they provide. For instance, insects are critically important as food for a great many other living things, including us. (We'll be addressing their role in human diets in a separate chapter.) Insects are abundant (often the greatest share of animal biomass in many habitats) and highly nutritious. Virtually all birds, apart from some seabirds, eat insects at one time or another during their lifecycle, and some are exclusively insectivorous. Flycatchers, swallows and martins, swifts (Figs. 13.9-12), and bee-eaters are some of those that are exclusive insectivores, but quail and falcons (Figs. 13.13-15) also occasionally rely on insects. Insects also feed many reptiles and amphibians, fish, and, of course, other arthropods like spiders, centipedes, and other insects (Fig. 13.14-16).

**Figure 13.9.** Eastern Kingbird, a flycatcher.

**Figure 13.10.** Barn swallow.

Good Guys and Goodies: Insect Products and Services

**Figure 13.11.** Purple martin.

**Figure 13.12.** Common swift.

**Figure 13.13.** Bobwhite quail. Quail chicks must consume large quantities of insects as they grow.

**Figure 13.14.** Peregrine falcon, a famous aerial predator of birds that also eats large flying insects.

**Figure 13.15.** A jumping spider (that's my wedding ring in the background).

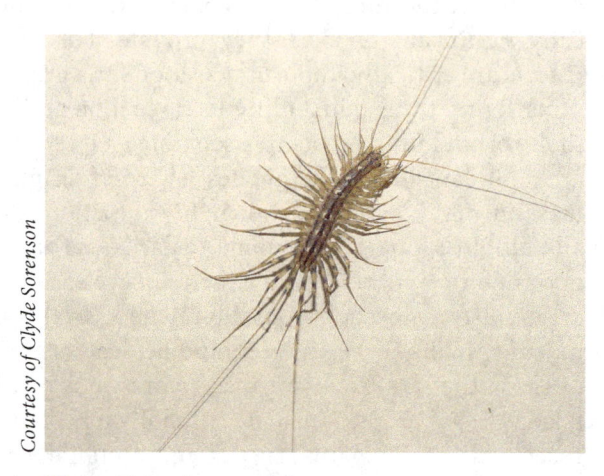

**Figure 13.16.** House centipede.

118   Insects and People

Insects also feed some plants! Carnivorous plants have long fascinated people because of the obvious role reversal, the "man-bites-dog" oddity of a vegetable eating an animal. Carnivorous plants typically live in nutrient-poor environments where nitrogen and phosphorus, critical plant nutrients, are very scarce. Insects are rich in protein and hence nitrogen, and they also contain phosphorus, and, as we said earlier, they are often abundant. Plants have evolved a number of odd and effective ways to capture insects and other small animals. Pitcher plants (Fig. 13.17) have leaves that form tubular vessels, often with a "lid" that prevents excess rainwater from flooding them. The leaf secretes water and digestive enzymes into this vessel. The walls and lip of the pitcher are slippery due to the presence of mucilage glands, and the upper walls of the pitcher often have downward pointing trichomes ("hairs"). Insects are attracted to the pitcher by sweet odors, land on the lip of the pitcher, slip, and fall into it, and then can't escape because of the slick, smooth walls and hairs. They eventually drown and are digested by the enzymes in the liquid. One can often get a unique picture of the arthropods in an area inhabited by pitcher plants by opening one of the old leaves and examining the hollow carcasses of the many insects it killed. Perhaps not surprisingly, some arthropods and other animals have evolved to exploit the unique habitat of a pitcher plant pitcher. There are mosquitoes whose larvae are found only in pitchers, and spiders that steal prey from the plant!

**Figure 13.17.** Yellow pitcher plant trap leaves.

**Figure 13.18.** Pink sundew.

Sundews (Fig. 13.18) are basically nature's flypaper. Their leaves are covered with long trichomes tipped by glands that secrete a sweet, sticky substance. Insects that land on the leaves (in response to attractive odors and colors) get stuck on the trichomes, and, slowly, over many hours, the leaf may fold itself more or less around the trapped insect. Digestive enzymes harvest the useful nutrients over many days, and then the leaf slowly unfolds to catch another victim.

**Figure 13.19.** Venus flytraps.

The most famous of carnivorous plants is, of course, the Venus flytrap (Fig. 13.19). This plant fascinated Charles Darwin. He considered it the most remarkable plant in the world. The flytrap is found in the wild only in the fire-maintained savannahs of southeastern North Carolina and a small area in northeastern South Carolina, and it is famous because of the "snap-trap" leaves it possesses. The trap consists of two oval lobes with long, curved spines along the margins. The inner part of the lobe is usually reddish in color, and there are three small hairs arranged in a triangle near the center of each. Contrary to popular conception, the traps don't clap shut. Unsprung, the pads are convex when viewed from the interior (the rims are flexed out relative

**Figure 13.20.** A closed flytrap pad with a captured grasshopper. Note the convex exterior shape and concave interior shape of the closed pads.

**Figure 13.21.** Salvia flowers displaying "landing pads" for potential pollinators.

| Table 13.2 | *A few Insect Pollinated Crops* |
|---|---|
| Alfalfa | Strawberries |
| Eggplant | Avocado |
| Asparagus | Stone fruits (peaches, plums, cherries, etc.) |
| Squash | Citrus (oranges, etc.) |
| Almonds | Apples |
| Cucumbers | Melons |
| Blueberries | Cacao (chocolate) |
| Beans | Cabbage, broccoli, and other cole crops |

to the center). When at least two of those six trigger hairs are touched within 20 seconds, there is a sudden change in leaf turgor that causes the rims to pop in, making the pad's interior concave (Fig. 13.20) and allowing the spines to mesh, entrapping appropriately-sized insects. Over the next day or so the pads close tightly over the insect and the cavity is flooded with digestive enzymes; once the nutrients have been extracted, the pads open once again, ready to trap another victim. Sadly, the Venus flytrap is threatened in the wild, due to habitat loss and illegal over-collection.

# No Insects, No Chocolate (and Lots of Other Food)

Another important service insects provide is pollination. Pollination is the process of transporting the male germ cells of plants to the female germ cells of other plants, and while a number of agents can accomplish pollination, the two most widespread are insects and wind. If a plant has apparent, showy flowers, it is most likely pollinated by some kind of insect (although some are pollinated by birds and bats). Among the crop plants we rely on for our sustenance, almost all fruits and vegetables are insect pollinated (grain crops like wheat, rice, and maize are all wind-pollinated). Table 13.2 lists some of these insect pollinated crops.

Plants attract pollinating insects by offering a nectar reward for the service, and they advertise this reward through showy, conspicuous flowers. Many plants have also evolved mechanisms to ensure that visiting pollinators have visited other plants of the same species, since pollen from a different species is useless to them. They do this by producing flowers that favor one kind of pollinating insect over others. Some plants in the mint family have flowers that act basically as levers; they don't expose the nectaries unless an insect of the right weight and size alights upon the "landing pad" (Fig. 13.21). Others have unique corolla shapes that only allow insects with mouthparts of the right shape. However, some plants and some insects cheat; many orchids, for instance, advertise a nectar reward without actually providing one, and some bumblebees steal nectar by chewing through the side of a flower. Some orchids even mimic the female of certain wasps both in appearance and odor, and are pollinated by males attempting to mate with them!

# In the End, We Are All Just Bug Food

Their ability to exploit and decompose organic matter is still another important ecological service insects provide. In a great many cases, insects are the ultimate recyclers, feeding on dead organisms or the waste generated by living organisms and returning the nutrients trapped in these things to the environment. Termites are widely reviled because of the damage they do to our wooden structures, but they are critically important to recycling wood and other plant debris in most tropical, subtropical, and temperate ecosystems. A host of other insects, including a tremendous diversity of beetles, contribute to this function.

**Figure 13.22.** African dung beetles.

Animal dung is another substance insects exploit. One animal's waste is indeed another's gold. Animals that feed on the waste of other animals are called **coprophages**. Even though we find dung repugnant, it does contain an abundance of potential nutrients for those creatures that have evolved to exploit it. It not only contains bits of food that pass unmolested through the defecator's digestive tract, it also contains large quantities of bacterial cells and the products of bacterial digestion that are highly nutritious to the right coprophage's digestive tract. Prominent among those insects that utilize dung are dung beetles (Fig. 13.22) and dung flies (Fig. 13.23). In some systems, these "dungivores" remove essentially all dung from the soil surface within hours of its deposition. We have exploited the ability of insects to clean up animal waste by introducing, accidently and deliberately, Old World species of dung beetles, which specialize in the dung of Old World livestock, to parts of the world where these livestock are exotic.

**Figure 13.23.** Mating golden dung flies.

The dead bodies of animals, large and small, are yet another resource utilized by insects. Animals that feed on the dead bodies of other animals are called **necrophages**. If it is warm enough, insects are invariably the first members of the clean-up crew to discover and exploit an animal carcass. Usually, the very first are certain species of blowflies. These insects can detect the earliest scents of death, and often find a carcass within minutes of the animal's death. There is a relatively predictable succession of insects that colonize a carcass; the blowflies are followed by fleshflies, which are succeeded by rove beetles, which are followed by dermestid beetles, and so on. Left alone, and in the right environmental conditions, insects can reduce a rather large vertebrate to literally skin and bones in a matter of days.

We exploit necrophages in two rather direct ways, beyond appreciating their mortuary service for all the animals that die every day. The fact that insects exhibit ecological succession on a dead animal, together with the observation that insect development is temperature dependent and predictable, allows insects to be used as important tools in forensic science. This is a special branch of entomology called **Forensic Entomology**. In this specialty, insects are used to provide evidence at crime scenes. We can use insect succession and development to help establish how long someone has been dead, or whether the body was moved from where the murder occurred. This kind of evidence requires very careful science to back it up, but if an investigation is conducted correctly, and all the necessary data is available, there is no better way to estimate the postmortem time following a murder if it is longer than a few days. Insects also have other forensic applications. Fragments of insects can suggest the geographic origin of contraband. Blood contained in a mosquito collected at the scene of a crime could potentially identify someone who had been there.

We also exploit necrophages through a medical application. The larvae of most species of necrophagous flies feed only on dead tissue, and we can utilize this in debriding ("cleaning up") wounds that are difficult to treat through conventional surgery, such as bedsores, or the circulation lesions the develop in folks with severe diabetes. We call insects used in this application "**surgical maggots**." This is actually an ancient practice that was used in Rome and rediscovered during the American Civil War. The larvae are raised in antiseptic conditions, and the sterile insects are applied, in a containing bandage, to the wound. The larvae remove the dead tissue without harming the healthy tissue with much more precision than even the most skilled surgeon. They also secrete an antibiotic chemical called allantoin that reduces the chance of infection. Wounds treated with maggots often heal faster and more completely than those treated with conventional surgery.

Insects have also been critically important to science. They are, in many ways, the ideal "guinea pig" for scientific investigations. Insects have short generation times, are extremely fecund, take up little space compared to other research animals, and can be reared much more inexpensively than other potential research subjects. Perhaps most important, insects, even though they look and act quite different from us, are animals with which we have very much in common. The humble pomace fly, *Drosophila melanogaster* (Fig. 13.24), has been essential to the development of our understanding of genetics, and it is still the workhorse of genetics research. This and other species of insects have been used to advance our understanding of animal behavior, ecology, animal physiology, and toxicology, among many other disciplines. Insects, of course, are economically and medically extremely important, and so a tremendous volume of research has been and continues to be conducted on the various ways we interact with them and they with us.

**Figure 13.24.** *Drosophila melanogaster*, the workhorse of genetics research.

Finally, insects have tremendous impacts on our cultural environment, on our literature, art, and entertainment. We'll save this topic for a later chapter.

# XIV

# Buzzers, Biters, and Stingers: Arthropods That Cause Direct Injury

Years ago, I spent a very long and uncomfortable night in a rustic cabin on the South Core Banks in Cape Lookout national seashore. My discomfort was initiated by a smoky brown cockroach that ran across my foot as I was just about to nod off. It was prolonged by another that bounced off my reading light and fell into my bedding, and it was completed by the hordes of no-see-ums that forced me to wrap myself in my sheets like a mummy to keep them from biting me in my sleep. Even for entomologists, insects can be annoying and aggravating. While the vast majority of insects are completely harmless to humans and should be regarded as beneficial creatures, there are a number of insects and other arthropods that are capable of negatively affecting us in one way or another. These injuries range from emotional responses to, occasionally, some very grievous physical assaults.

A great many folks find various and sundry insects annoying in some way. Some are distressed by the appearance of certain insects. Their odd shapes, "hairy" aspect and general conformation make creatures like carpet beetle larvae (Fig. 14.1) or many flies aesthetically undesirable. Other folks are repulsed by the behaviors of some insects. A roach scurrying across the floor when the kitchen light is turned on is alarming; a camel cricket (Fig. 14.2) leaping in your direction when you open the crawlspace door can be intimidating (although the creature is completely harmless). Others are bothered by the odors some insects produce. Few people find the aroma of aggravated kudzu bugs (Fig. 14.3) pleasant. Still others find the sounds insect make disturbing. While dozens of crickets and katydids pleasantly chirping away outside your bedroom window may lull you to sleep, one single cricket chirping inside your bedroom might keep you up all night.

**Figure 14.1.** Carpet beetle.

**Figure 14.2.** Camel cricket.

**Figure 14.3.** Bean plataspid, aka the "kudzu bug."

In general, how annoying you find various insects is largely dictated by your experiences with insects as you grow up. People who grow up in rural areas of the South have a much different experience than folks who grow up in an urban center in the Northeast, and their attitudes towards insects will be radically different. Generally, people are more tolerant of creatures they encounter frequently and less tolerant of the exceptional. People who grow up in other parts of the world have even more dramatically different tolerances. Many Australians are much more tolerant of flies buzzing around their heads than most North Americans because of the presence of the Australian bush fly, a close relative of the housefly, which prefers to land on the faces and heads of large animals like us. One is typically taught attitudes towards insects by parents and the environment.

While almost everyone finds some insects annoying, some people deal with more severe emotional responses to insects. **Entomophobia** is a generalized fear of some or all insects, often whether they actually represent a threat or not. This condition can arise from a bad experience with insects or from a lack of experience with, and knowledge of, insects. Entomophobia is a very real problem for those who suffer from it, but it can be treated very effectively with a number of therapies. As in many similar afflictions, knowledge can be a powerful tool for overcoming fear.

A more severe emotional condition involving arthropods is **delusional parasitosis**. In this condition, sufferers are convinced that they are infested with creatures, often living in their skin, and that this infestation is seriously compromising their well-being. Upon examination, however, no parasites can be found, and any sores the victim exhibits are self-inflicted wounds made in an attempt to remove the non-existent creatures. Delusional parasitiosis is often a symptom of more severe mental or emotional illness, and the prognosis for many sufferers is not very good. Delusional parasitosis often is also associated with a history of actual infestation, or proximity to an actual infestation. The recent resurgence of bed bugs in the United States has led to an increase in this syndrome, but, while the possibility of arthropods might be a trigger for the condition, the actual cause is usually something much deeper.

# Biters and Stingers: Arthropods that Cause Physical Injury

Many, many things eat insects, as we established in the previous chapter, and, as a consequence, most insects have defensive tools at their disposal to help protect them from their natural enemies. Many others deliberately seek us out for the resources we contain. In this section, we'll discuss the insects and other arthropods that can inflict actual, direct injury to us.

Caterpillars are essentially, at the most fundamental level, moving bags of protein and fat, intent on eating as much as they can to make more protein and fat. They are, for the most part, slow and soft, and they have fairly limited perception of the environment around them. Many caterpillars therefore fend off potential predators with setae or spines modified to cause pain and irritation in attackers. **Urticating** setae are stiff, barbed structures that break off readily when they brush against an attacker. They embed in the skin and cause sometimes severe, though usually short-lived, itching. Some caterpillars are armed with more potent, hollow spines connected to venom-secreting glands at their base. Brushing against these injects the venom into the skin, causing more severe and persistent itching and burning. One of the most common "stinging" caterpillars in North America is the saddleback (Fig. 14.4). This distinctive caterpillar has a wide host range, and its stinging spines produce an intense burning sensation that lasts for an hour or so (in my experience, anyway). The puss moth caterpillar (Fig. 14.5), however, produces a much more severe reaction. The short stinging spines are hidden under the long, silky, non-stinging setae. Puss moth envenomation can produce body-wide systemic symptoms, and the victim may be debilitated for several days. There are stinging caterpillars in tropical South America that produce venom potentially lethal to humans.

A substantial number of insects have large, sharp mandibles that they use to subdue prey or to chop food into manageable pieces, and many of these can inflict a painful defensive bite if provoked. Native cultures have actually exploited some of these insects for medical purposes. The soldiers of some of the larger driver, army, and leaf-cutter ants have been used as sutures to close wounds (Fig. 14.6).

Many insects have non-venomous spines and spurs that can potentially puncture human skin, but, with most of these creatures, the human "victim" has to ask for it by mishandling the insect. The biggest mantids can penetrate human skin with the large spines on their raptorial forelegs (Fig. 14.7), and some of the big stick insects have large spines on their hind legs, powered by very strong leg muscles, that they drive into attackers (Fig. 14.8).

**Figure 14.4.** Saddleback caterpillar, an insect well-defended by urticating spines.

**Figure 14.5.** The puss moth caterpillar, perhaps the most dangerous North American stinging caterpillar.

**Figure 14.6.** Burchell's army ant soldier.

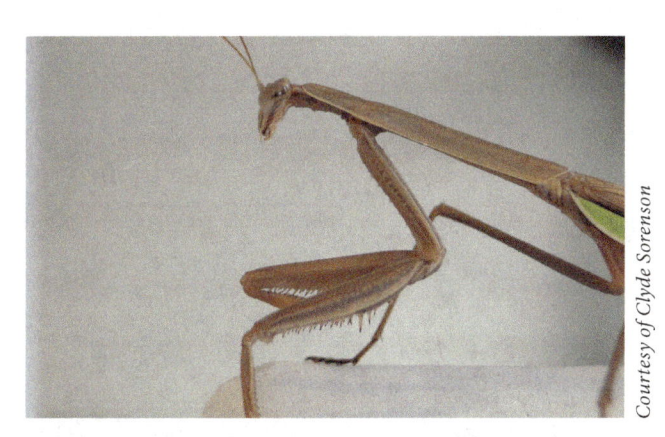

**Figure 14.7.** The heavily armed prothoracic legs of a Chinese mantid.

**Figure 14.8.** Goliath stick insect.

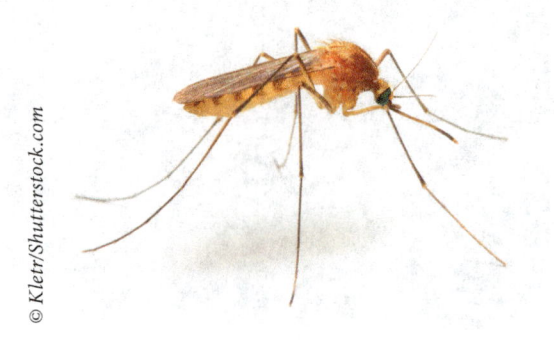

**Figure 14.9.** Mosquito.

**Figure 14.10.** Horsefly.

**Figure 14.11.** Deerfly or "yellowfly."

An unfortunately large number of arthropods, particularly a diversity of true flies (order Diptera) seek us out for something we have—our blood. To many of these animals, we are basically big moving pools of protein and other nutrients. Most of the biting species of flies need the protein to make eggs, and so, in most biting flies (with some important exceptions) only the females bite us. This is true of mosquitoes, deerflies, horseflies, blackflies and no-see-ums or punkies (Figs. 14.9–13); most of these insects inhabit relatively nutrient-poor aquatic habitats as larvae and need the additional nutrients to successfully reproduce. This blood piracy is risky business, so different biting flies have different biting strategies. Mosquitoes are stealth biters. They try to slip in unnoticed, alight lightly, and inject anesthetics with hypodermic needle-like mouthparts to steal your blood without your noticing. Horseflies, on the other hand, are hit-and-run biters. They zip in, flay your skin open with mouthparts that look very much like a steak knife (Fig.14. 14), and lap up as much blood as they can before you try to kill them. Biting flies also play an important role in the transmission of diseases, which will be covered in another chapter.

A number of other arthropods seek us for our blood or other body fluids. The blood-sucking conenoses or kissing bugs (Fig. 14.15) are a group of South and Central American assassin bugs in the order Hemiptera that have abandoned the predaceous lifestyle of their ancestors for one as parasites of vertebrates. The species that are significant problems for humans live in the rafters, roofing, and other harborage in human homes and descend to the sleeping areas at night to get their blood meals. They often bite around the mouths of their victims, hence the name, "kissing bug." Both sexes and all life stages are blood feeders, and they are important vectors of the increasingly important Chagas disease, an infectious disease that left untreated can cause congestive heart failure.

**Figure 14.12.** Blackfly.

**Figure 14.13.** Biting midge, aka "punkie" or "no-see-um."

Insects and People

**Figure 14.14.** Horsefly mouthparts.

**Figure 14.15.** Eastern blood-sucking conenose.

Chiggers (Fig. 14.16) are the bane of southern blackberry pickers and others who love the outdoors. Also called redbugs, they are the almost microscopic larvae of a mite that is not parasitic in the adult stage. They are most common in the weedy, brushy areas where their preferred hosts, rodents and small birds, are likely to be most abundant. An unlucky and imprudent person could end up with hundreds of bites from one brief visit to chigger habitat. Chiggers bite and then inject saliva, which lyses skin cells. The saliva and the material released from the cells forms a hardened tube through which the chigger feeds. The intense itching one experiences in a chigger bite is an allergic response to the foreign proteins in this tube.

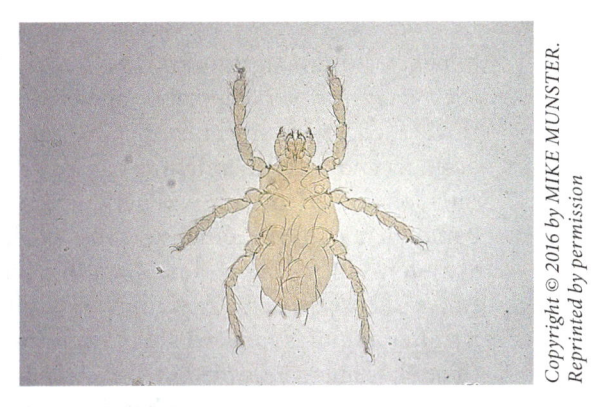

**Figure 14.16.** A larval harvest mite, or chigger.

Humans aren't ideal hosts for chiggers, and they often either drop off or die within a few hours, but they leave that itch-inducing tube behind. Chiggers typically settle where they perceive constriction, such as under the elastic bands of socks or undergarments, or in places where two body parts press together, as in the armpit. In general, they like to settle in places that are not polite to scratch in public. Chigger bites can be treated with topical analgesics and oral antihistamines, but the best strategy is to avoid getting them in the first place. The same tactics that protect one against tick bites will also protect against chiggers, and we'll be covering that topic in the next chapter.

## Venomous Fangs

Several arthropods are armed with venom glands that they use to subdue prey, but that also can be used defensively when the animal is threatened. All centipedes, for example, are venomous, although not all have venom claws ("fangs") suitable for envenomating humans. Centipede size can be deceptive. While most of the large species can bite, some of the smaller ones can, too, as I learned the hard way a few years ago when a two-inch specimen tagged me on my pinky finger. While several North American species can bite, no species are terribly dangerous. The bites are about as painful as a wasp sting (more on them in a moment), and the symptoms last about the same length of time.

Perhaps the most notorious arthropods, at least in terms of venomous bites, are the spiders. While all spiders are technically venomous, only a relative handful of species are capable of biting humans, and, of these, only a very small number are medically significant. In the United States, the two types of spiders that cause the greatest concern are the widows and the recluses.

**Figure 14.17.** Female black widow spider.

There are three species of black widow spiders in North America, and at least one can be found pretty much everywhere south of southern Canada. All three species are quite similar in appearance (Fig. 14.17). The females are medium-sized, shiny black spiders with a rather bulbous abdomen marked underneath with a red spot or spots that often, but not always, assume an "hourglass" pattern. The males are much smaller and more colorful, with abundant red and white markings on their bodies and legs. Black widows can be very common. They typically build their rather untidy, cob-web type webs in secluded spots under debris, in crawl spaces, in woodpiles, and in other protected, out-of-the-way spaces. Only adult females pose any significant threat, since only they have "fangs" long enough to penetrate human skin, and they generally will only bite when under duress. They may bite when they accidently get trapped as one is moving material around, or when their silken egg cases are threatened, and many bites are "dry" and don't actually inject any venom. They are not aggressive animals and will generally try to retreat if given the chance. Black widow venom is neurotoxic. If one does experience envenomation, it results in an extremely unpleasant experience. One major component of the venom results in a widespread release of neurotransmitters, resulting in systemic symptoms including profuse sweating; body-wide, severe muscle cramps; racing pulse; and sometimes intense, body-wide pain. The symptoms usually subside within a day or two, but their severity often requires medical attention to manage them. Deaths to widow spiders are extremely rare (none have been recorded in the United States in several decades), but they do occur. The closely related brown widow spider has been introduced to the United States and appears to be expanding its range. It is somewhat smaller and its bite appears to be significantly less severe with symptoms more localized than those of black widows.

**Figure 14.18.** Brown recluse.

For many people, the recluse spiders are even more frightening than black widows, probably because they are less well known and the reputation of the consequence of a bite has been somewhat mythologized. In the United States, the most common of these spiders is the brown recluse (Fig. 14.18), which is native to a large area of the mid-southern United States. Outside of this area, this is an extremely rare spider. A great many other spiders are frequently misidentified as brown recluses, and, indeed, untold millions of wolf spiders and other completely harmless brown spiders are dispatched annually for being "recluses." The defining characteristics of the true recluse are a uniformly fawn-brown body marked by a fiddle-shaped, darker brown marking on the top of the cephalothorax; unmarked, fawn-brown legs; and six eyes rather than the eight most spiders have. The brown recluse, as its name suggests, is an extremely retiring creature that prefers dark and undisturbed areas. It is a prowling spider that actively hunts its prey, rather than relying on a web to snare it. Like the widow spiders, they are not at all aggressive and only bite in response to direct threat or entrapment. People in the Mid-West frequently live for years with hundreds of brown recluses in their houses and never experience a bite. Most bites occur when someone puts on clothing harboring a recluse, or traps one with his hand while moving materials harboring the spider.

Brown recluses are notorious because the venom they produce is necrotizing. It kills tissue, often resulting in a large, sunken and permanent scar at the site of the bite. Treatment for recluse bites is mainly directed at managing other symptoms and at keeping the wound clean until it (slowly) heals. However, brown recluse bites may be one of the most common misdiagnoses in modern medicine. Many other insults to the body can produce lesions similar to those produced by recluse envenomation, including infections from a number of bacteria that may be introduced to small cuts, scratches, or other small wounds. Most "recluse bites," even in areas where the spiders are fairly common, are probably actually nasty bacterial infections of such minor wounds.

While there are several other North American spiders that can bite humans, most are extremely reluctant to do so, and bites of these other spiders generally produce much, much milder symptoms than the widows and recluses (Fig. 14.19).

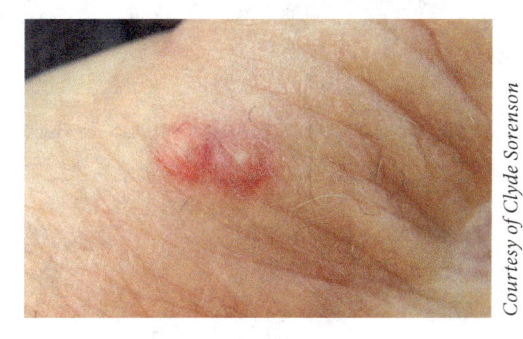

**Figure 14.19.** My thumb, showing the relatively mild (but still annoying!) symptoms of a bite from a yellow sac spider.

## Ladies to Beware of: Stinging Hymenoptera

The Hymenoptera, the bees, ants and wasps, have well-deserved reputations for being able to defend themselves. The stinger of a bee or wasp is, of course, a modified ovipositor, and, since only females have ovipositors, only females can sting. The accessory glands associated with this modified ovipositor have assumed the role of venom glands, and in many social Hymenoptera, the venom they produce has evolved to help the insect deal with potential nest predators—like us. Hymenopteran venoms vary dramatically in their effects on humans. Some, like that of sweat bees (Fig. 14.20), elicit very little discomfort, while others like that of paper wasps (Fig. 14.21) and velvet ants (Figs. 14.22), which are actually large wingless wasps, can be extremely painful. Some stings, like that of the tarantula hawk wasps of the desert Southwest, or the bullet ants of tropical Central America (Fig. 14.23), can be completely debilitating. Justin Schmidt, an entomologist from Arizona,

**Figure 14.20.** A sweat bee in a Venus flytrap flower.

**Figure 14.21.** A paper wasp.

**Figure 14.22.** A female velvet ant (pinned specimen).

Buzzers, Biters, and Stingers: Arthropods That Cause Direct Injury    129

**Figure 14.23.** A bullet ant, purported to have most painful sting in the world.

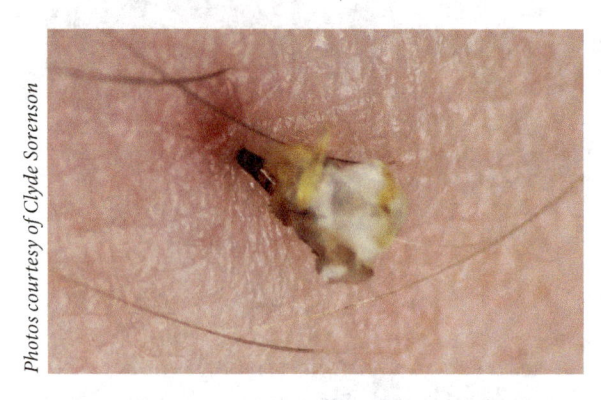

**Figure 14.24.** Honeybee stinger and venom glands. A. Stinger in wound; note reddening skin; b. stinger removed. Note barbs on stinger and now-depleted venom glands.

attempted to establish a pain scale for Hymenoptera stings by experiencing them himself and then writing comparative descriptions.

Hymenopteran venoms are usually complex mixtures of several chemicals that together produce the painful experience after the sting. Honey bee venom, for instance, contains at least nine different chemicals; some cause local cell destruction, while others increase pulse rate and dilate local blood vessels, increasing the spread of the venom, while still another helps initiate an itchy allergic response. The efficacy of the venoms of social Hymenoptera is heightened by the effects of the alarm pheromones found in many of these species. The alarm pheromone rallies other members of the colony to the attack, so that the invader receives multiple stings, and presumably, a stronger message to move on and leave the colony alone. If a person can't retreat from the colony's attack, this can be a very dangerous situation since in some cases in sufficient quantities, the venom of many of these insects can be acutely toxic and potentially fatal to humans.

Most stinging Hymenoptera have smooth ovipositors and can sting multiple times if one allows them to. Fire ants and yellow jackets will latch onto the skin of an invader with their mandibles, and then sting multiple times in a semi-circle, using their fixed jaws for leverage. Worker honeybees, however, have barbed stingers and can only sting once. After embedding the stinger in the attacker's flesh, a bee rips the end of her abdomen off and then dies shortly after. The stinger, with its venom glands, stays behind (Fig. 14.24) and muscles continue to pump venom into the wound for many seconds unless removed.

# You're Never Alone: Arthropods that Parasitize Humans

Compared to insects, we are absolutely immense beings, and to some, we are all, or almost all, the habitat they need. Parasites are organisms that live on other organisms (the hosts), generally at the hosts' expense, and a number of arthropods are parasitic on humans for all or part of their lifecycles.

Prominent among these creatures are the three forms of true lice that are found on humans. Two of these forms, the head louse and the body louse, are actually different populations of the same species, *Pediclus humanus* (Fig. 14.25). The closest living relative of this species is a louse found on chimpanzees. The head louse is restricted to the human head and feeds on the scalp of the infested person. The complete lifecycle takes about two weeks. Females glue their eggs, or nits, to the shafts of hairs at the scalp, and, as the hair grows, the nit rises above the scalp. The eggs hatch in about two to three days, and it takes about nine days to go through the three nymphal stages. All nymphs and adults must have access to human blood through the scalp, and they can't live off the host for more than a day or two. Head lice aren't known to transmit serious diseases.

Head lice are not an indicator of poor hygiene or socio-economic status. They are merely an indication that you have been around someone who has lice. The only way to contract head lice is to have head-to-head contact with someone who has them or to use their toiletry items (brushes, etc.) or infested bedding. The overwhelming majority of cases of head lice occur in children, particularly girls, under the age of 11 or so (although their parents often end up infested, too). In North America, folks of Caucasian ancestry are far more likely to get head lice than those of African ancestry because the most prevalent lice populations in North America have adapted to hair that is round in cross-section (and hence tends to be relatively straight). In Africa, lice have adapted to the oval cross section of the very curly hair most folks of African ancestry have. Infestations can be controlled through the judicious use of lice shampoos containing certain insecticides, lice combing to remove the nits, and laundering or otherwise treating potentially infested personal belongings. Continued vigilance for a few weeks is necessary after the infestation is cleared to ensure that they don't flare back up or recolonize.

Body lice appear to have diverged from head lice about 70,000 years ago, about the time, we think, people began to wear clothing of one sort or another. Body lice actually live in, and lay their eggs in, clothing, and periodically (about five times a day) take a short journey to the skin to get a blood meal. Since they rely on this particular habitat, body lice generally can't persist on people who have regular access to hygiene and who routinely launder their clothing. Body lice are, therefore, usually associated with people who are experiencing severe socio-economic distress. During times of war or other calamity, when people are often forced to live in close proximity and with few resources, body lice populations can build to huge levels, often with disastrous consequences for the afflicted populations, as we'll see in a later chapter. (Unlike head lice, body lice are the vectors for several very serious diseases.) In modern North America, they are most often found on homeless people who don't have the opportunity to regularly bathe and wash their clothes, and who often live in crowded shelters or makeshift housing. Lice bites itch intensely due to an allergic response to the anesthetics and anticoagulants they inject when they bite, and this is important in disease transmission. Individuals can also harbor thousands of lice, biting thousands of times a day. One feels "lousy" not just from the itching, but also because of the gross blood loss! Body lice can be controlled with the reestablishment of routine hygiene and the sanitation of any infested belongings.

**Figure 14.25.** Human body louse.

The third form found on humans is the crab, or pubic, louse (Fig. 14.26). This louse is adapted to large diameter, coarse hair and is most common in the pubic region, but can also be found in other areas, including the chest hair of men, the armpits, and facial hair including beards, mustaches, and eyebrows. The pubic louse, like the head louse, doesn't transmit disease and is itself transmitted primarily through sexual contact with infested persons. The crab louse is most closely related to a louse found on gorillas.

**Figure 14.26.** Pubic louse.

Several species of mites are also intimate parasites of humans. Scabies is a skin condition caused by a nearly microscopic mite (Fig. 14.27) that actually burrows through the upper layers of the skin. Afflicted skin areas often have dry, bite-like spots in lines, a scaly texture, and they itch intensely. Scabies can only be transmitted from person to person through extended skin-on-skin contact, and the condition is most prevalent in situations where folks are forced to live in crowded and relatively unhygienic conditions. In the United States, they are most commonly found among the tenants in nursing facilities and among the homeless, but even in these populations, they are relatively rare.

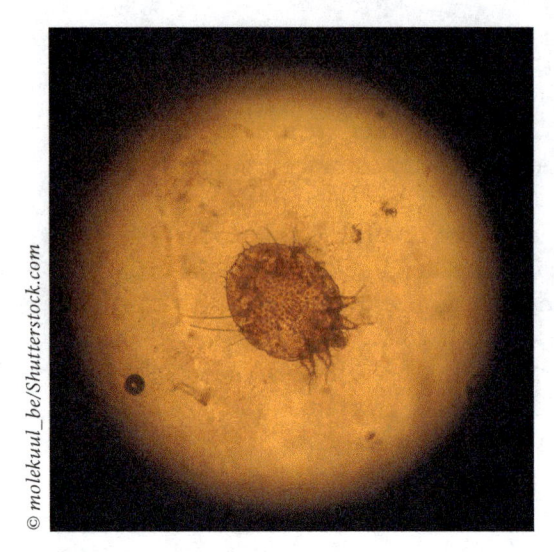

**Figure 14.27.** Human scabies mite.

Another kind of mite, however, is much, much, more common—in fact it is quite likely that you harbor these creatures as you read this page. These are the two species of *Demodex* follicle mites (Fig. 14.28). These creatures aren't actually parasitic; they feed on the secretions produced in the follicles of eyelashes and eyebrows, and on the sebum and its degradation products produced in the sebaceous glands on your face. For most of us, they are completely harmless, although they may cause acne in some people and more severe symptoms in very few. The older one gets, the more like he or she is to have a population. One acquires his population of eyelash mites by head-to-head contact with someone who already has them, and many folks get them from their mothers, when they are children. They reproduce in the hair follicles and reach adulthood in about two weeks. While it is possible to rid one's self of them, for most of us, it's probably not worth the effort.

**Figure 14.28.** Human follicle mite.

The larvae of some flies are also parasitic on humans. We call the condition where fly larvae inhabit living, vertebrate animal tissue "myiasis." One of the most notorious of these insects is the human botfly, which is native to Central and northern South America. The human botfly has one of the more bizarre life cycles among insects. The adult female fly captures a mosquito and lays an egg on it. She then releases the mosquito, which flies off to find a blood meal. When the mosquito finds a suitable host, the egg hatches in response to the body heat of the victim, crawls down the mosquito's leg, and tunnels into the site of the mosquito bite. The larva then develops a warble, a small cell in the skin in which it lives and feeds. It also establishes a hole in the skin through which it breathes. After eight weeks and three larval stages, the creature crawls out through the breathing hole and drops to the ground where it pupates. Adults don't feed—they just hunt mosquitoes (and sometimes other flies) upon which to oviposit. Full-grown larvae approach an inch in length (Fig. 14.29), and sustaining one is a fairly painful experience, particularly at night when the larvae seem to be most active. The warbles rarely become infected since the larvae secrete antibiotic compounds into the wound. Disgusting as the experience is, it is relatively easy to remove the larva. Swabbing the wound in

**Figure 14.29.** Larvae of the human botfly.

petroleum jelly suffocates the larva, which can then easily be extracted with forceps. Insect repellents and tight-weave, long-sleeved clothing can help prevent acquisition of a botfly larva when one travels in botfly country.

A few insects are parasitic on us but only hop on board when they need a meal. Most folks are familiar with fleas (Fig. 14.30). Fleas are holometabolous, wingless relatives of scorpionflies. All are parasitic as adults, while the larvae of most species feed primarily on the feces of adult fleas in the nests of the host animals. Far and away the most common pest flea around human habitations in North America is the cat flea. While this species does indeed prefer cats, it is quite content to feed on many other kinds of animals, including us. Both male and female fleas bite. The key to avoiding flea bites is managing the fleas on our pets, and, in recent years, we have developed a great many tools to facilitate this task.

Perhaps a bigger concern for most people is the resurgence of an old pest, the bed bug (Fig. 14.31). The bed bug is an obligate parasite on humans and the animals that live with them. Their closest relatives are species that inhabit the nests of swallows and other cliff- and cave-dwelling birds. These insects live in the cracks and crevices of beds and bedrooms, and slip out after dark to sip blood meals from us as we sleep. Both adults and nymphs are hematophagous, and while they so far have not been implicated in transmitting diseases, they can cause severe allergic reactions that can be extremely discomfiting to many. For most of our history as a country, bed bugs were a fact of life; everyone encountered them at some point or another, and most folks lived with them. However, the development of DDT as an insecticide changed everything, and bedbugs virtually disappeared from the United States. Of course, there were tremendous problems with DDT, and it disappeared from the scene in the 1970s. By then, the insects had so completely vanished that most folks not only didn't have any experience with them—they never thought of them, except perhaps when they used the familiar bedtime farewell.

**Figure 14.30.** Human flea.

In the late 1980s, however, bed bugs started to reappear. Most current populations are resistant to some of the insecticides that replaced DDT, and with our highly efficient transportation system and globetrotting ways, we've managed to transport them all over. Bed bugs are now a huge problem in many US cities, particularly in hotels and in apartment complexes. While bed bug problems can be eliminated, it is costly and difficult. The best approach to bed bugs is to not get them in the first place. It is a very good idea to carefully inspect any hotel room you may be renting before moving in. Place your luggage in the bathtub, and then look under the mattress, box spring, and headboard for any evidence of the critters (the most obvious signs are the dark droppings the insects leave in their harborage).

**Figure 14.31.** Bed bug.

# The Greatest Danger: Arthropods and Allergy

For most people, the consequences of negative interactions with arthropods are minor and transient, but for some, particular arthropods, that they may routinely encounter, may present grave threats to their health and even their lives, by instigating allergic reactions. Allergy to insects spans a wide range of responses.

Allergy is, simply put, an overreaction of one's immune system. In a normally functioning immune system, introduction of a pathogen to the body elicits the development of antibodies to antigens on that pathogen. The antibodies identify the pathogen and the immune system destroys the invading organisms and any cells that harbor it. In allergy, the immune system comes to recognize otherwise harmless substances as dangerous. The range of symptoms generated by this overreaction can range from mild, i.e., congestion and a scratchy throat, through hives and intense itching, to the most severe reaction, anaphylaxis, where global bodily immune responses can kill the victim.

Chronic exposure to insect fragments or to their waste can eventually sensitize people to the point that they develop debilitating allergic responses. This is true for me. After years of working with different moths that are important agricultural pests, I have developed an allergy to the scales from lepidopteran wings. If I aspirate just a few, my sinuses clog up and my throat gets dry. An antihistamine generally relieves my symptoms, and on I go about my business. However, many folks live in an environment where exposure to insect related allergens is constant, and far more severe reactions, such as asthma, become commonplace. Perhaps the best example of this kind of syndrome involves the German cockroach, which is far and away the most common domestic pest roach in North America. Research conducted over the last decade or so has conclusively established that a majority of cases of asthma in inner city, urban neighborhoods is due to chronic exposure to the epicuticular products from German cockroach exoskeletons.

Another arthropod that is frequently implicated in allergy is the house dust mite (Fig. 14.32). This creature is virtually ubiquitous in our homes. They are particularly abundant in and around our beds, where they feed on bits of organic debris, prominently the millions of skin cells we shed every day. The shed exoskeletons and feces of the dust mite are potent allergens for some people, and it is extremely difficult, if not impossible, to completely eliminate them from a dwelling.

Anaphylactic shock resulting from bee, ant, and wasp stings is the most significant threat insects pose, but only to those who are severely allergic to them. While these venoms are toxic, for most people, it takes a very large number of stings to deliver a dose of venom large enough to seriously threaten life or health, since each sting delivers only a tiny amount of venom. However, for folks who are allergic to a particular hymenopteran venom, a single sting is potentially enough to kill. One sting can initiate a cascade of immune responses that result in plummeting blood pressure, dizziness and loss of consciousness, swelling of the extremities, and life-threatening swelling of the throat and trachea. As we stated earlier, all venoms are not alike, and allergy to one kind of venom doesn't necessarily mean that someone will have a severe reaction to the sting of another species (although there are cases of cross-reactivity). In the United States, 50-100 people die each year from Hymenopteran stings, and virtually all of these are due to severe allergic reactions.

**Figure 14.32.** House dust mites in a pillow.

# XV

# Guess Who's Coming to Dinner!
## Insects as Food

For most modern North Americans, the idea of eating an insect is inherently disgusting. You, as you read this, are probably experiencing a rather visceral reaction to the idea at the mere mention of it. But for most of our time as a species, and, indeed, for most of the people currently living on our beautiful little blue planet, the idea of eating insects was and is no more alarming than the idea of eating a hamburger is for us. **Entomophagy** (eating insects) is an ancient and proud human tradition and may be something North Americans might want to reconsider.

We have probably always eaten insects. Our closest living relatives, the two species of chimpanzees, relish six-legged snacks and have developed tool-using cultures to harvest them. Some cave and rock art appears to depict arthropods, including those that might be edible. Many of the relic hunter-gather societies that remained until the last century made abundant use of insects and other arthropods for food. Many tribes of Native Americans also ate insects. The Paiutes of the Great Basin (what is now Nevada and parts of Oregon and Utah) made tremendous use of insects. During Mormon cricket outbreaks, they harvested the insects by the tens of thousands, dried them, and made flour that they stored for hard times. Since the development of the written word, prominent authors have discussed insects as comestibles. Aristotle, in his *History of Animals*, described the insects he liked to eat, and Pliny discussed how grasshoppers might be fattened and made more delectable by feeding them flour and wine. Even the Bible has an entomophagy section. In Leviticus there are detailed instructions describing which insects a good Jewish person of the time could eat (those that hop) and could not eat (pretty much all the rest).

Eating insects makes good evolutionary sense for an omnivorous species like us. They are extremely nutritious (more about that in a moment), often extraordinarily abundant, and, it turns out, not at all bad tasting. We in North America relish many other foods that, on the surface might strike some as odd, and some of these appetites are relatively recent developments. We eat beef, which is abhorrent to many south Asians. Many of us have acquired a taste for raw fish over the last few decades (an idea my dad was never able to wrap his mind around), and most of us eat insect relatives (shrimp, crabs, and lobsters) with gusto. Eating insects is a foreign idea to us not because it's a bad idea, but because this practice is not part of our northern European cultural heritage.

It's important to realize that in many places and at many times, insects have represented the single largest available source of animal protein. Remember, as we said early on, insects grossly outnumber as well as outweigh us. Prior to the emergence of agriculture, "plagues" of locusts and other insects were probably considered gifts from the gods.

**Figure 15.1.** Coddling moth caterpillar in an apple.

The bakers among you are probably familiar with the practice of sifting flour prior to measuring it out for a specific recipe. While we currently do this to eliminate clumps and standardize the density of the flour, originally, the main point was to sift out the big bug bits before you made the batter. Indeed, before we developed technologies for enriching the vitamin content of flour, the insect fragments in flour made a substantial contribution to the micronutritional value of it. Insects are so ubiquitous that most of us eat significant amounts of insect all the time, perhaps two to three pounds over the course of a year. Insect contamination of food is virtually impossible to eliminate, and, indeed, the US Department of Agriculture acknowledges this fact by publishing guidelines for legally allowable amounts of insect and other contaminants (Table 15.1). Insects and fragments of insects are found in fresh fruits and vegetables, and in processed products like flour and chocolate. For the most part, this contamination has no negative effect whatsoever on human health or the utility of the food. Insect contamination only becomes a food safety issue when it reaches high enough levels to allow other organisms, like fungi, to gain a foothold (Fig. 15.1), or when the contamination consists of insects that are harboring, or potentially harboring, disease-causing pathogens.

Our aversion to eating insects does cause one major problem, in that attempts to produce insect-free food result in otherwise unnecessary pesticide use. In many food production systems, farmers can have entire truckloads of fruits or vegetables condemned by a processor for minimal insect contamination because the processor is not willing to take on the risk of lawsuits by insect-averse consumers who find what they consider to be unacceptable insect contamination. A grasshopper incidentally incorporated into a can of collard greens poses no threat whatsoever to the consumer who finds it there, yet, in one such case, a woman in Virginia was awarded $2,000 dollars for the headaches and depression she experienced following the discovery. An accumulation of such lawsuits can damage a processor's reputation, and so, the processor demands "clean" produce. We have to decide if "clean" is defined by complete lack of insects or by reduced insecticide residues.

| Table 15.1 | Maximum Allowable Insect Contamination of Various Foods (source: U.S.D.A) |
|---|---|
| Canned tomatoes | 2 maggots/ 17.5 oz. can |
| Raisins | 35 fruit fly eggs/ 8 oz. box |
| Mushrooms | 20 fungus gnat maggots/3.5 oz. can |
| Whole peppers | 1% insect infested on average |
| Canned apricots | 2% insect infested or damaged |
| Orange juice | 10 fruit fly eggs or 2 larvae/ 9 oz. can |
| Tomato sauce | 30 fruit fly eggs or 15 eggs and 1 larva per 3.5 oz. can |
| Sesame seeds | 5% infested |
| Peanuts | 20 insects per 100 lbs |
| Canned black-eyed peas | 5 weevil larvae/ can |
| Frozen broccoli | 60 aphids/ 3.5 oz. |
| Beer hops | 2,500 aphids/ 0.35 oz. |

The observation that a great many other animals rely on insects suggests that they might be nutritious. Insects are indeed very high-quality food, since they are high in protein and in desirable types of fat, and compare quite favorably with some other sources of protein and fat (Table 15.2). They are also high in vitamins, minerals, and other micronutrients, although the exact amounts of these nutrients vary with different insect species. Some species are particularly high in fat at certain times of their lifecycles; for instance, reproductive, alate termites leave their colony (and are harvested by humans) when they are carrying all the energy reserves in the form of fats they will need to establish new colonies.

Of course, not all insects are universally edible, since a great many contain toxins or have armaments meant to protect them from predators (like us). In general, insects that are very brightly colored in contrasting patterns of red or yellow with black or white (Fig. 15.2) should be avoided, since these aposematic, or warning, patterns may indicate chemical toxin protection. Similarly, insects armed with urticating spines, dense, hairy setae, or stingers (Fig. 15.3) should probably be avoided as well (although some of these are routinely eaten after appropriate processing). Most entomophages (people who eat insects) also tend to avoid adult beetles and other insects with very hard exoskeletons, because the ratio of nutrient to cuticle is too low to make it worthwhile, and, probably, because the crunch is too much! Many of these same species, however, are extremely palatable and nutritious as larvae and pupae.

While perhaps 80% of the human population at least occasionally eats insects on purpose, entomophagy is deeply entrenched in some cultures. Honey bee larvae and pupae, mayfly and caddisfly immatures, silkworm pupae, grasshoppers, and cicadas are all relished and

**Figure 15.2.** A monarch butterfly caterpillar. The bright, aposematic coloring is a warning that it is toxic.

**Figure 15.3.** A milkweed tussock moth caterpillar protected by, and made unpalatable by, dense setae.

### Table 15.2  *Nutritional Value of Insects Compared to Other Meats (per 100 g serving)*

| Species | Calories | Protein (g) | Iron (mg) | Thiamine (mg) | Riboflavin (mg) | Niacine (mg) |
|---|---|---|---|---|---|---|
| Termite | 613 | 14.2 | 0.75 | 0.13 | 1.15 | 0.95 |
| Caterpillar | 370 | 28.2 | 35.5 | 3.67 | 1.91 | 5.2 |
| Palm weevil | 562 | 6.7 | 13.1 | 3.02 | 2.24 | 7.8 |
| Lean ground beef | 219 | 27.4 | 3.5 | 0.09 | 0.23 | 6.0 |
| Broiled codfish | 170 | 28.5 | 1.0 | 0.08 | 0.11 | 3.0 |

**Figure 15.4.** Fried giant waterbugs.

**Figure 15.5.** Mopane worms (*Gonimbrasia zambesina* caterpillars) —an African delicacy.

are prepared in a wide range of styles. Thai folks enjoy honey bee immatures and a wide range of other insects. In Thailand and Vietnam, a species of giant waterbug (Fig. 15.4) is particularly valued for a "spice" derived from its pheromone glands. The whole insects are also enjoyed fried or incorporated into chili pastes. Markets across different parts of Mexico may feature as many as 200 different species of insects that are regular parts of the diet for folks living in these regions. In South Africa, a very significant cottage industry has developed around the harvest, preparation, and sale of mopane worms (Fig. 15.5), the caterpillars of one of the giant silkworm moths. In parts of Columbia, fried or roasted leaf-cutter ant queens are a seasonal delicacy. Insects are such a regular part of the diet in many of these cultures that canned insects from these traditions can be found at ethnic food markets in the United States.

World-wide, approximately 1,400 species of insects and other terrestrial arthropods are commercially significant as food in at least a limited way, although only a handful of the 30 or so orders of insects contain significant numbers of these species. In terms of total global poundage, Orthoptera is probably the most important food insect order. Grasshoppers and crickets are large, mostly phytophagous insects that are abundant and relatively easy to collect. There are also commercial firms that raise large numbers of, in particular, crickets. Other orders eaten in large quantities are the Coleoptera (beetles, mostly the larvae), Lepidoptera (moths and butterflies, mostly the caterpillars), Hymenoptera (bees, ants, and wasps, mostly the larvae and pupae), and Isoptera (mostly alate reproductives). Many other insects from other orders are eaten as well.

Insects are extremely versatile when it comes to preparation, with some caveats. Unless they have been preserved through canning or drying, only live or freshly killed insects should be eaten. Insects are small animals with large surface areas, and, like shrimp and crabs, they deteriorate very rapidly after death unless measures are taken to prevent decay. Also, one should only eat insects of known provenance; purchase them from a reputable insectary ("insect farm"), or collect them yourself. Once in hand, insects can be dried and ground to make a nutritious flour, roasted and seasoned to make a snack food, deep fried (breaded or not), or incorporated into soups or stews. Basically, they can be prepared pretty much the same as any other protein.

Different insects, as one might expect, have different flavors, much as different fish or shellfish do. In general, insects taste much like what they eat: corn earworms (the large colorful caterpillars you may have encountered when shucking sweet corn) taste, not surprisingly, like sweet corn; mealworms (the larvae of a grain-feeding beetle) taste much like cornbread when fried. Many ants have a sour or vinegary flavor, thanks to the formic acid found in their venom glands, while others taste vaguely of pine or juniper. Most bee, ant, and wasp larvae are very mild in flavor (the overall impression is one of buttery smoothness), and they typically take on the tastes of accompanying ingredients. Most aquatic insects are reminiscent of fish or shrimp.

While entomophagy is not a common practice in North America, it is growing. For most of our recent history, insects have been eaten for shock value, but over the last decade or so, serious entomophagy has become more common. There are a number of prominent restaurants in major cities that now include insects on their menu at least occasionally, and insect cooking contests have become a regular part of some major educational events like the Insecta Café at the North Carolina Museum of Natural Science's annual Bugfest celebration in Raleigh. There have even been several cookbooks published to encourage the practice. The biggest impediment to increased entomophagy in the United States may be sourcing the insects for cooking. We don't have infrastructure built around producing commercially viable quantities of insects for human consumption, so it is difficult to find them, and they are expensive when one does. However, as entomophagy gains traction in this country, this infrastructure will develop, and within the last few years, several companies have begun producing insect products for the food industry, including cricket flour that is being used in some snack nutrition bars. If you get the chance, try eating some insects—you'll probably be surprised by how good they are.

# XVI

# Tick, Tack, Sick: Tick-Vectored Diseases and What to Do About Them

Ticks are among the most aggravating arthropods that feed on us, and they can be among the most dangerous. Many years ago, in an earlier existence, I worked on a project describing the distribution and population size of the American alligator in North Carolina. One of the tasks we did was to search for alligator nests in the wetlands around Lake Ellis-Simon and along the tidal rivers around Wilmington. This meant slogging through marsh and swamp vegetation for many hours a day, looking for the large mounds of decomposing plants the creatures use to incubate their eggs. Typically, when we left the swamp, we'd have dozens to hundreds of ticks on us, which often required hours of effort to remove.

Ticks are, of course, not insects but rather are arachnids. They are basically large, parasitic mites, which must take several blood meals from vertebrates in order to complete their lifecycles. With each blood meal, they have the opportunity to pick up disease-causing microorganisms. It is their role as transmitters of disease that makes them so potentially dangerous.

In order to discuss diseases that are transmitted by insects and other arthropods like ticks, we need to define some important terms. A **disease** is a negative response in a living organism to some entity that could be living or non-living, or even endogenous (generated by the organism's body itself). A good example might be West Nile encephalitis (brain inflammation). An **agent** is a foreign "organism" that causes disease in another organism. In West Nile disease, for instance, the agent is a virus (viruses aren't exactly living organisms) (Fig. 16.1). A **vector** is the living organism that transmits the agent from an infected host to an uninfected host. In West Nile disease, the vectors are several mosquitoes (Fig. 16.2). The vectors may or may not experience negative effects

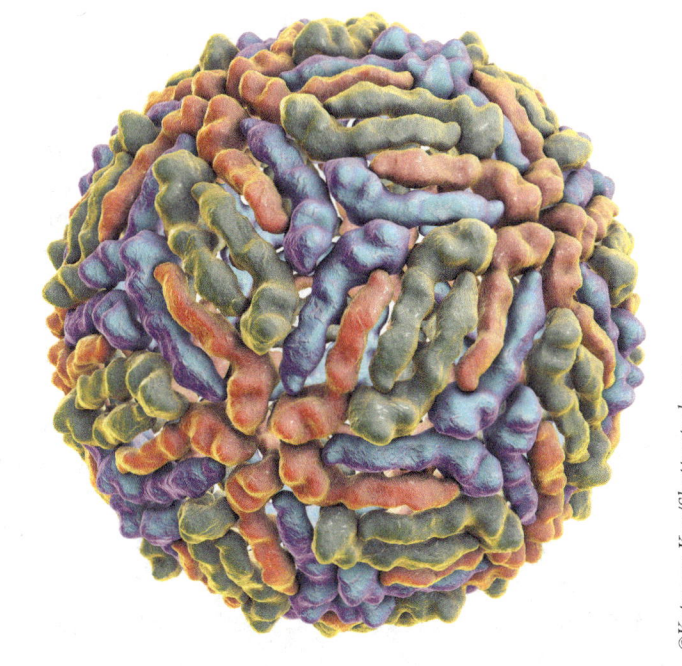

**Figure 16.1.** An illustration of a West Nile viral particle.

from carrying the agent. A **reservoir** is a living organism that maintains the disease agent in the landscape over long periods of time. In West Nile disease, the reservoirs for the virus are birds, mostly songbirds. In most arthropod vectored diseases, the reservoirs maintain the agent with little or no ill effects to themselves, although there are many exceptions to this rule.

Vectors may transmit disease agents in several ways that fall into two major categories. Some agents are **transmitted mechanically**. In these diseases, the agent basically "hitches a ride" on the tarsi (feet) or mouthparts of the vectoring arthropod. Remember from our discussion of mouthparts in Chapter 4 that houseflies more or less automatically drop their mouthparts and regurgitate a bit of their last meal on whatever they land upon. A housefly feeding on, say, an eight-day-old, unrefrigerated cheeseburger (or something far, far nastier…), and then alighting on your fresh ham-and-Swiss could potentially mechanically transmit disease agents from that bad food to your good food. Similarly, a cockroach that runs through "stuff" in the bathroom and then across a kitchen counter could move pathogenic agents. Indeed, filth flies, like the housefly, and cockroaches have both been implicated in the transmission of a host of protozoan pathogens.

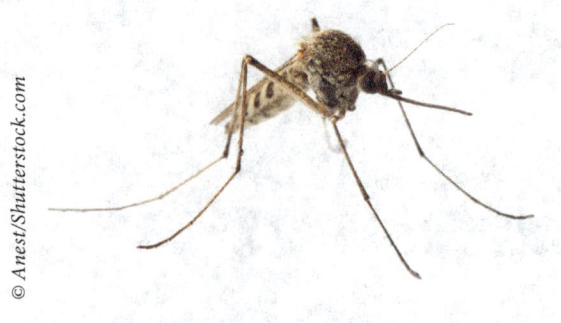

**Figure 16.2.** A *Culex* mosquito.

Most arthropod-vectored agents, however, are **biologically transmitted**. In biological transmission, the agent spends some time inside the vector, and, in most, must pass some critical portion of its lifecycle inside the vector. At the very least, it replicates inside the vector. African sleeping sickness, for example, caused by a trypanosome protozoan, goes through two transformations inside the tsetse flies that vector it in order to be transmitted.

While a number of human diseases are vectored by arthropods in North America, those ticks I mentioned earlier transmit some of the most important. Three of these are particularly prominent in various regions of the continent.

## Lyme Disease

In the early 1970s, epidemiologists noted an unusual concentration of pediatric arthritis in the vicinity of Lyme, Connecticut. Arthritis is generally quite rare in children, so this concentration was a huge red flag that something odd was going on. After several years of brilliant medical sleuthing, researchers identified a spiral-shaped bacteria as the cause of the illnesses, the spirochete *Borrelia burgdorferi* (Fig. 16.3), and that the black-legged deer tick (Fig. 16.4) transmits it. We've also learned that the reservoirs for this disease are deer mice and white-tailed deer, at least in eastern North America. Since its discovery, this disease has been recognized in most states of the Union, although by far the greatest concentration of cases is in New England, with a smaller concentration in the western Great Lakes states. Lyme disease is certainly the most commonly reported tick-borne disease in the United States.

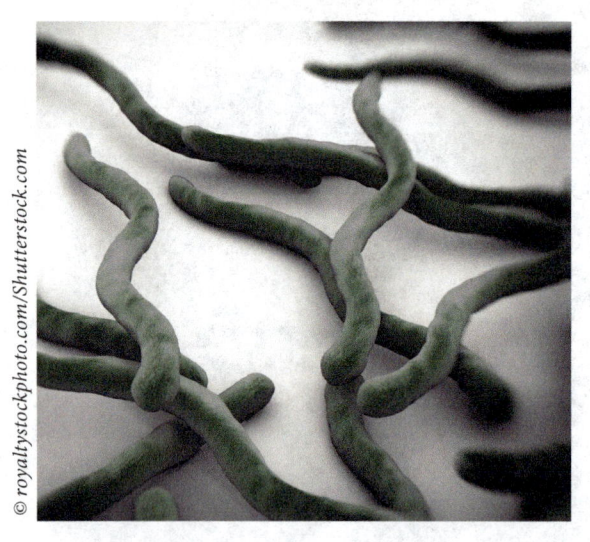

**Figure 16.3.** An artist's conception of *Borrelia burgdorferi*, the spirochete agent responsible for Lyme disease.

142   Insects and People

Black-legged deer ticks are three host hard ticks, meaning that they must take three blood meals, from three different hosts, to complete their lifecycle. ("Hard tick" means they belong to a group of ticks with a hard, protective shield, or scutellum, protecting their unengorged bodies.) Females lay masses of eggs that hatch into six-legged larvae. Larvae seek small mammals, like deer mice, and birds for their first meal. If the mouse they bite harbors the spirochete, they pick it up and are capable of transmitting it in later stages. After feeding to repletion, the larva drops off and molts into the eight-legged nymph, which seeks a blood meal on a rodent or, perhaps, a deer or human. After again feeding to repletion, it drops off and molts to the adult stage, which seeks a blood meal from a large mammal like a deer, dog, or, again human. Mating occurs on the host. The males host seek primarily to find females. After the mated female feeds to repletion, she drops off and eventually lays one large batch of eggs, after which she dies. It usually takes two years to complete this lifecycle. Unlike many other three host ticks, black-legged ticks are cool season ticks and are most active when temperatures are moderate, rather than in the heat of the summer.

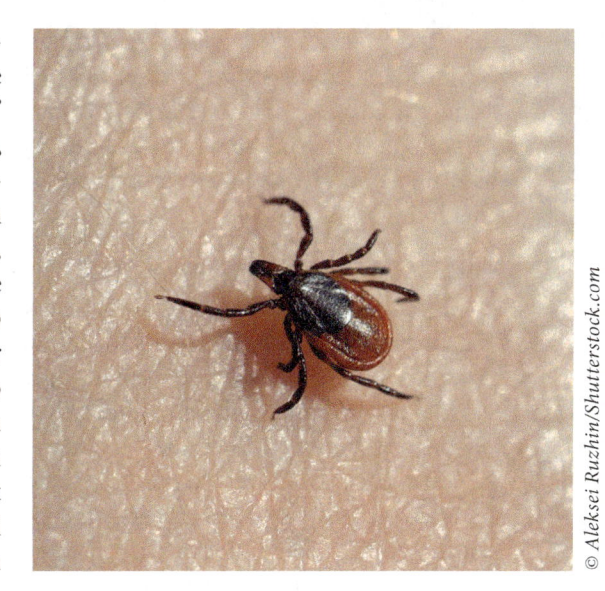

**Figure 16.4.** *Ixodes scapularis*, the black-legged deer tick, vector of Lyme disease.

The symptoms of Lyme disease can, unfortunately, be mistaken for many other diseases, and vice versa, at least initially. Most sufferers experience general malaise, not unlike what one experiences with the flu, including fever, muscle and joint aches, headaches, and nausea. These symptoms usually start within two weeks of the infective tick bite. In between 50 and 80% of cases, a characteristic "bulls-eye" rash develops in the vicinity of the bite. The actual bite site will be red and often raised, and this will be surrounded by a ring of clear skin, which is then surrounded by a ring of reddened skin. While the actual bite site may be quite itchy, the outer ring won't generally be. The red ring actually represents the front of bacterial proliferation. If Lyme disease is left untreated, more severe consequences can develop, including permanent damage to joints, cardiac complications, and in some cases, neurological injury and brain damage. If diagnosed promptly, however, it is very readily treated with appropriate antibiotics.

Lyme disease appears to be less common in many southern states, including North Carolina, for a couple of important reasons. First, the bacterium is simply less common in the reservoir population, which alone greatly reduces the possibility of transmission to humans. Secondly, the larvae of southern black-legged ticks prefer to feed on lizards and other reptiles, reducing their opportunity to pick up the pathogenic agent. All that said, Lyme disease does occur in these states and can't be ignored.

# Rocky Mountain Spotted Fever

In the summer of 1986, I spent a great deal of time hopping around in roadside ditches, checking moth pheromone traps I deployed as part of my Ph.D. research. Somewhere near the middle of that time, I came down with what initially felt like a bad cold, but with a fever. The day after my symptoms set in, I experienced the worst headache of my life, before or since, and knew that something was seriously wrong. I went to the student health center, where a very old, but very wise doctor asked me if I had experienced any tick bites in the last couple of weeks, and, of course, I had. "Did any stay attached for any length of time?" he asked, and, of course, one had. "You're an entomology student, right? What kind of tick was it?" I replied, "American dog tick." "Ah hah!" he said. "You, my lad, probably have Rocky Mountain Spotted Fever. Take these antibiotics according to the directions, and you'll be fine. And be more careful about ticks!" With that, I was dismissed, and, as it turns out, I was fine after a few days. Without treatment, however, it could have gone much, much different.

Rocky Mountain Spotted Fever (RMSF) is caused by another kind of bacteria called *Rickettsia rickettsii* (named in honor of the guy who discovered it and who died of a closely related organism). It was first recognized in the Rocky Mountains of the American West, but it occurs over much of North America. Indeed, the highest number of cases year in and year out occur in the southern states, and North Carolina and Oklahoma typically lead the nation in cases. In the southern United States, the primary vector for RMSF is the American dog tick (Fig. 16.5), which is another three host tick and the largest species in this region that routinely bites humans. Fortunately for us, typically only the adults bite people, making them rather conspicuous compared to some other ticks. In other parts of the country, other related species are responsible for transmitting the bacterium. The reservoir for RMSF is a range of medium-sized wild animals and the ticks themselves. Unlike some of the other diseases we'll talk about, the agent for RMSF can be transmitted from the female tick to her eggs, through a process called transovariole transmission.

**Figure 16.5.** The American dog tick, *Dermacentor variablis*, an important vector of Rocky Mountain Spotted Fever.

RMSF is a very dangerous disease. Left untreated, it frequently kills people, but as in my case, prompt administration of appropriate antibiotics leads to rapid and complete recovery. I exhibited most of the classic early symptoms; as the disease progresses, illness gets more severe and debilitating. In many, but not all cases, a characteristic rash develops. The rash exhibits as a pinpoint, red rash that starts at the wrists and ankles and spreads towards the trunk. Usually, by the time the rash appears, the patient is in serious medical trouble. Prior to the development of effective tetracycline-type antibiotics, about 25% of the people who contracted RMSF died of it. Many of the survivors were left with permanent kidney, heart, or brain damage. With modern antibiotics, however, no one should suffer such a fate—as long as treatment is swift.

## Ehrlichiosis

Ehrlichiosis is a relatively recently described disease that is widespread through the United States, particularly in the South. The agents that cause this disease are three species of *Ehrlichia* bacteria, which all produce very similar disease symptoms. The primary vector of the *Ehrlichia* pathogens is the lone star tick (Fig. 16.6), which is one of the most common ticks in the southern United States. The reservoir for these pathogens appears to be deer.

Lone star ticks are aggressive biters of humans in all three stages. Many folks traditionally call the tiny, abundant, and extremely avid larvae and nymphs, "seed ticks." The adult male is slightly smaller than the female pictured in Fig. 16.6, and they lack the prominent white spot in the middle of the back. Lone star ticks, like the others we've talked about, are three host ticks. Like black-legged deer ticks, but unlike American dog ticks, lone stars are "deep biting" ticks. They have long mouthparts, which they insert relatively deeply into the skin. Many people also find their bites far more irritating and itchy than the bites of some others.

Ehrlichiosis shares many symptoms with the other diseases we've discussed so far, except that in most people, there is no characteristic rash. These pathogens attack white blood cells, and in rare cases can cause death. Like the other diseases, ehrlichiosis can be treated very effectively with appropriate antibiotics. This disease is probably more under-reported than some of the others, and they, too are underreported.

In addition to these three diseases, ticks are capable of transmitting many other diseases to both humans and domestic animals. It's important to remember that your pets are susceptible to tick-borne diseases, too. Dogs, for instance, can contract Lyme disease.

**Figure 16.6.** Female lone star ticks (*Amblyomma americanum*). The lone star tick vectors the agents responsible for Ehrlichiosis, among other diseases.

Ticks can also be responsible for a couple of non-infectious diseases. An American dog tick bite can cause a syndrome called tick paralysis. It is most commonly associated with ticks attached on the head or neck. In this illness, a reaction to a toxin in the tick's bite can, over days, result in complete paralysis and severe breathing problems. Sufferers recover quickly once the tick is removed.

A second, relatively newly recognized malady is colloquially called "alpha-gal" allergy. This is an allergic reaction to the sugar galactose-*alpha*-1,3-galactose, which is found in the tissues of all mammals other than Old World monkeys and apes (we are, technically, apes). This allergy develops when one is exposed to the sugar through the bites of lone star ticks that have previously fed on other mammals. Once a person has been sensitized to the sugar following tick bites, eating red meat can trigger an allergic response. People afflicted with this allergy may have to give up red meat for the rest of their lives, all though there is some evidence that the allergic response fades with time as long as another tick bite can be avoided. Sufferers are able to eat fish, shellfish, and poultry with no ill consequence, however, since the sugar is not found in these animals.

## Protecting Yourself from Ticks

Ticks are a fact of life for folks who enjoy the outdoors, but with the proper precautions, it's easy to do what you want to do without fear of ticks and their diseases.

In order to board a host, a tick climbs to the top of a grass blade or weed stem, and extends its front legs out and up. On each of these legs is a structure called Haller's Organ, which enables the tick to sense the body heat and carbon dioxide produced by host animals. The tick may stay in this position for days or weeks until a suitable host walks by, but when one does, the tick latches on and rapidly crawls to a safe spot on the animal to settle and bite. Once the tick finds a refuge on the host, it takes quite some time for it to work its mouthparts into the skin and begin drawing blood into its gut. The mouthparts of a tick are barbed, rather like a harpoon, to securely fasten it to its host, and on top of this, the tick secretes a "glue" to make it even more difficult to remove. The tick also injects anticoagulants and anesthetics to ensure blood flows and to make sure the host isn't aware of it. It typically takes several days for the tick to fill to repletion, at which time it dissolves the glue, removes its mouthparts and drops to the ground. Because of the way ticks attach and feed, they generally have to be latched on for more than six to twelve hours before they actually start transmitting any pathogens they may harbor.

The first step in protecting yourself is to recognize tick habitat when you are in it. Ticks are generally going to be found where potential hosts will be found. Most of their hosts are animals of edges—the places where different habitats meet. Woods lines, brushy areas or meadows with tall grass are all likely spots, but dense woods can also harbor large numbers. If you know you will be in tick habitat, you then need to take the proper precautions.

The next step, then, is to dress appropriately. Wear long sleeves and long pants, ideally, light in color (to make it easier to see any ticks). Button collars and cuffs, and if you know you will be in particularly "ticky" habitat, tape your pants cuffs and sleeves. High rubber boots you can tuck your pants legs in are a good idea. Repellents are extremely useful, as well.

There are two major categories of repellents you may consider. The first is a clothing protectant called permethrin. Permethrin is actually a pyrethroid insecticide based on the chemical structure of a natural insecticide found in certain daisies. If properly applied to your field clothes and allowed to dry, it very effectively repels and kills any ticks (and chiggers!) that hop on board. It should never be applied to the skin, since it is rapidly absorbed and degraded (and can make some folks itch). One application can last through seven or eight launderings.

A large number of insect repellents formulated for application to the skin also can be useful. While most are marketed for protection against mosquitoes, some effectively repel ticks and chiggers. Products containing DEET have been around for quite a long time and do have good activity. However, many people find the chemical smell unpleasant, and it does have a tendency to "eat" plastic. Several newer products that rely on chemicals of botanical origin are also quite good. Products containing undecanone or geraniol are particularly effective. Application of one of these repellents to exposed skin, combined with permethrin clothing protectant, can provide virtually complete protection from ticks. I treat all my hunting and birding clothes with permethrin, and my skin with either a DEET or geraniol product, and I rarely ever find a tick after a day in the woods. (If only we had had permethrin back on the alligator project…)

A very important key to good tick disease prevention is to examine yourself frequently for ticks when you've been in tick country. Thoroughly check your skin and that of your companions (and your pets, as well). Bear in mind that ticks like to settle where skin is constricted by clothing, where skin meets skin, and along your hairline. However, they may settle anywhere, so complete examination is prudent. If you perform "tick checks" at least every six hours, it is highly unlikely any ticks will be attached long enough to successfully transmit disease.

Should you find a tick that has attached, there is one right way, and a host of wrong ways, to remove it. Contrary to popular belief, a tick can't be made to let go by attempted suffocation with petroleum jelly or nail polish, nor will cooking its backside with a recently extinguished match-head encourage it to release. Swabbing them with alcohol or kerosene is not effective either. Remember, the tick has glued a small harpoon in your skin—if you kill it, there's no way it can "let go." The only proper way to remove a tick is to grasp it as close to the skin as possible with a pair of forceps or tweezers and slowly, but steadily, pull until the mouthparts are extracted. In a pinch (sorry…), one's fingernails, protected with tissue, can also be successfully deployed. Very small ticks can be carefully scraped out with the sharp edge of a credit card or similar rigid implement. It's important to wash the bite site, your hands, and any tools you used thoroughly with soap and hot water as soon as you can after extraction. And don't discard the tick. Save it in alcohol with the date and location marked, or tape it with transparent tape to a calendar on the day you removed it, so that you have a record of the bite and the kind of tick that bit you, should you develop symptoms. Any suspicious illness in the following two weeks, any fever, muscle or joint aches or headache, should send you to seek medical attention immediately—and take your tick with you, so the doctor has some idea of which disease may be afflicting you.

# XVII

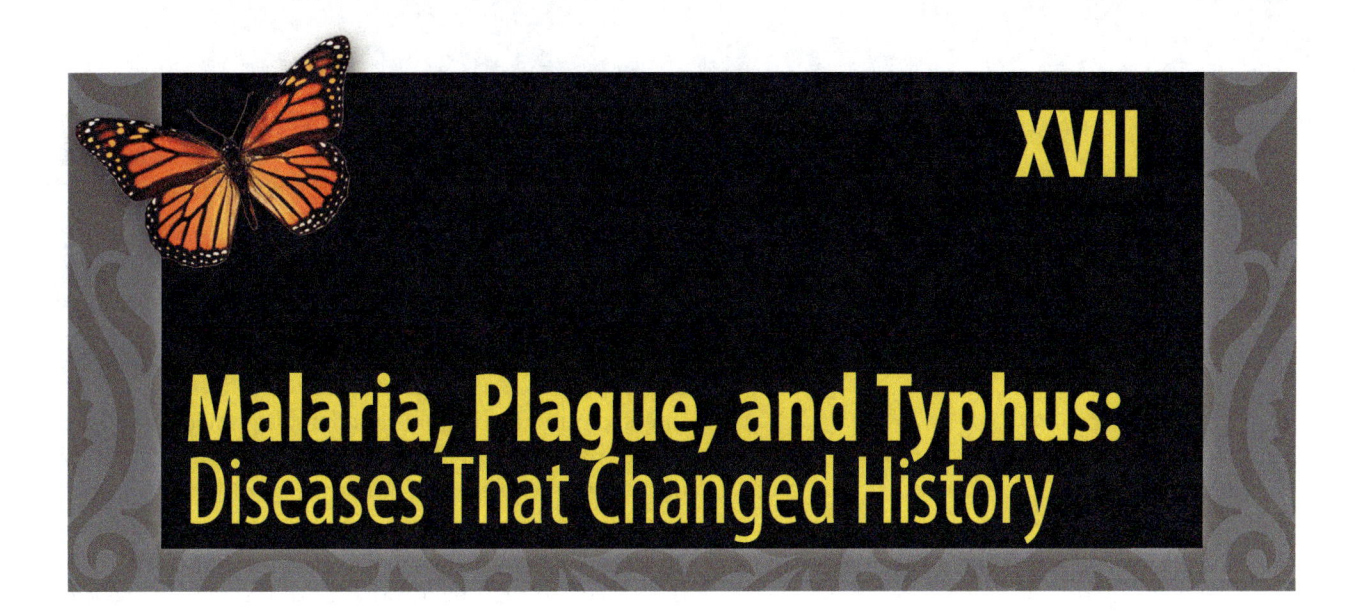

# Malaria, Plague, and Typhus: Diseases That Changed History

Some of the "natural history" channels on TV would have you believe that the most fearsome animals on the face of this beautiful, complex planet are the ones that, given the chance, could eat you—creatures like great white sharks, brown bears, lions, and the like. While I readily admit that disappearing down the gullet of any of these magnificent creatures is a fate greatly to be avoided, their toll through history pales in comparison to that exacted upon us by the ones with six legs.

Insects have undoubtedly killed vastly more humans than any other animals because of the diseases they transmit, and indeed, some of these diseases have altered the course of our evolution, as well as the course of our cultural history. Three diseases, in particular, stand out for the huge effects they have had on us (although many, many other diseases have had impacts as well). The most infamous of these is malaria.

## Bad Air: Malaria

Malaria (the word is Italian for "bad air") has always been with us. For most of our history, the cause of the disease was unknown, but early on, people made the connection between malaria and the swampy habitats of mosquitoes. The name malaria originated from this association, although folks thought the disease was contracted from breathing the "bad air" of the swamps. It is caused primarily by four species of *Plasmodium* protozoa (Fig. 17.1): *P. falciparum*, *P. vivax*, *P. malariae*, and *P. ovale*. A fifth species that primarily infects monkeys can, on rare occasions, cause disease in man. It's important at this point to note that there are a great many other species of *Plasmodium* that cause disease in other animals, including other primates and birds. *P. falciparum* is the most devastating of the four primary species that cause human disease, but infection with the other three species is by no means trivial.

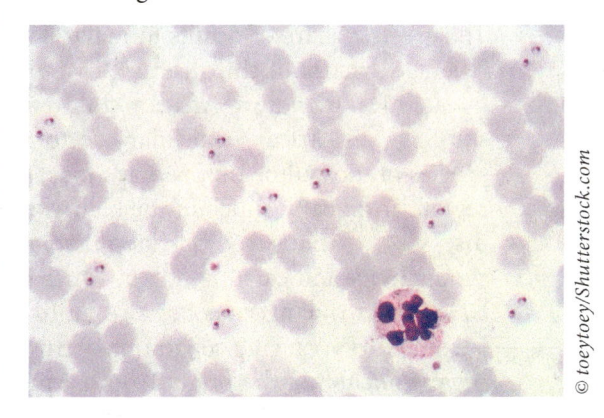

**Figure 17.1.** *Plasmodium* sp. malarial parasites in red blood cells.

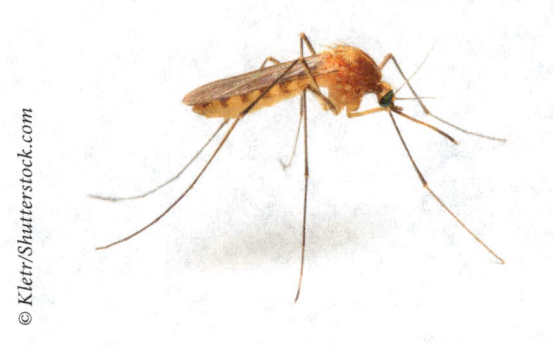

**Figure 17.2.** *Anopheles* sp. mosquito.

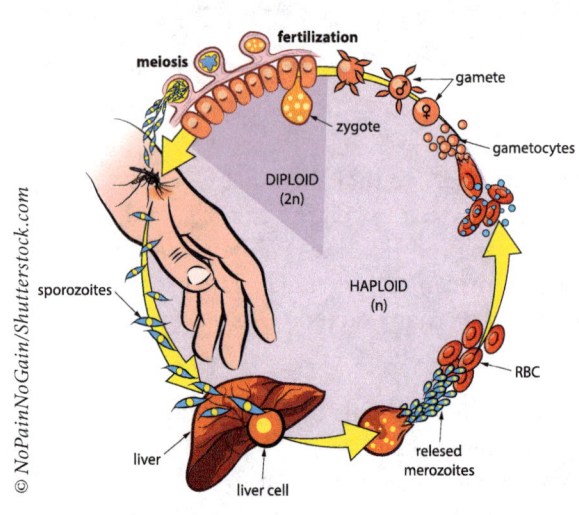

**Figure 17.3.** Malaria lifecycle.

Human malarial agents are vectored only by mosquitoes of the genus *Anopheles* (Fig. 17.2), which are found essentially world-wide. The only reservoirs for human malaria are humans. Most people who contract the disease don't die from it if they receive no medical treatment, but they will harbor the parasite for the rest of their lives and will experience chronic bouts of illness.

Malaria has a very complex life cycle (Fig. 17.3). A female mosquito picks up gametocytes while feeding on an infected human. The male and female gametocytes join in the insect's gut to form zygotes, which develop into oocytes in the gut wall. Sporozoites proliferate in these until they rupture. The sporozoites invade the salivary glands and then move to the next human host when the mosquito next feeds. The mosquito side of the life cycle takes about two weeks, depending on temperature, so the parasite relies on the mosquito living that long between blood meals. Once in the human bloodstream, the sporozoites migrate to the liver, where they hide from the immune system inside liver cells. Here, the protozoan undergoes another change, dividing and dividing again until hundreds of the resulting merozoites cause the infected liver cell to rupture, allowing them to invade the bloodstream. The merozoites invade red blood cells, where they reproduce, producing more merozoites to invade still more red blood cells. This is when the victim finally becomes ill. Some of the merozoites produce gametocytes, which get picked up by another mosquito to complete the cycle.

During the phases when it is in the bloodstream, the symptoms of malaria include severe fevers, chills, and profuse sweating, along with severe lethargy and general illness. These symptoms can reoccur every few days so that the victim becomes ill again just when he was beginning to feel better. Medications, which we'll discuss in more detail a bit later, can effectively clear the parasite. Two of the malaria species, *vivax* and *ovale*, also go into hiding in the liver cells and can cause new bouts of the disease ("relapses") years after it was first contracted and presumably cured.

The greatest share of deaths to malaria are caused by *falciparum*, and, if we were to describe the "average" victim of this disease, it would be a child under 5 living in Africa—the vast majority of fatalities are young African children. This is because *falciparum* is the prevalent species in sub-Saharan Africa; the primary vector in this area is a highly anthropophilic ("human-loving") species of mosquito, and, perhaps most importantly, extreme poverty is common in much of the region. (Ironically, much of the blame for this poverty can be laid squarely on malaria and other diseases—a vicious feedback loop that is extremely difficult to break. Diseases like malaria reduce productivity and earning power, which reduce the availability of resources to fight disease.) *Falciparum* is so dangerous because, unlike the other species, it causes infected red blood cells to become sticky, causing congestion of tiny blood vessels and sometimes catastrophic injury to major organs, including the brain. Approximately 200,000,000 people currently contract malaria each year. As recently as 15 years ago, malaria was killing as many as 2,000,000 million people annually, but thanks to renewed and concerted efforts by public health agencies and non-profit organizations, in recent years deaths have been reduced to approximately 500,000 or so.

Malaria has been such a long-entrenched part of our species' history that it has actually affected our genetics. Sickle-cell disease is a genetic disease that is incompletely dominant. If two heterozygous people (that is, each having one sickle-cell trait gene and one "normal" gene) marry, on average, one-fourth of their children will inherit two normal genes, one-half will be heterozygous, like their parents, and one-fourth will inherit two sickle-cell genes. These last, homozygous for the sickle-cell gene, will, without modern medical technology, often die before they are old enough to reproduce. Those who are homozygous for the sickling gene have red blood cells that are sticky and deformed (Fig. 17.4), particularly when the person is oxygen-stressed, and these deformed cells can clog small blood vessels, leading to organ and tissue damage. However, folks who are heterozygous for this trait have red blood cells that are somewhat resistant to malarial parasites, and so these people have some level of protection against the disease. That this genetic disease persists in populations exposed to malaria, even though it potentially kills one-quarter of the offspring of "protected" people, is a testimony to the evolutionary impact of this malady.

**Figure 17.4.** "Sickled" versus normal red blood cell.

Malaria has also had profound impacts on the course of our history. Prior to the 19th century in many parts of the world, malaria was almost a fact of life, and many prominent historical figures suffered from it. The disease may have killed both Genghis Khan and Alexander the Great, and Christopher Columbus and his crews were afflicted with malaria. *Falciparum* malaria was not native to the Americas but was introduced shortly after European colonization began. It had disastrous consequences for the native peoples who had no evolutionary experience with it. Eight US presidents from George Washington to John F. Kennedy, had malaria. In the southern and eastern United States malaria was endemic from colonization through the early 1950s, and many people alive today remember having the disease in their youth.

For most of history, a malaria infection was a storm one had to weather (unless you were Chinese—see below), but in the 17th century, Spanish colonizers in Peru noticed that a "fever bark" the Peruvian natives used to treat illnesses also worked to alleviate malaria. The bark comes from several species of cinchona trees. We call the active component of the bark, quinine. Quinine was the first medicine that effectively removed the parasites from the blood stream, and it proved a boon to malaria management. During the time of the British Raj in India, British military personnel were required to take a daily dose of quinine in the form of bitter tonic water to prevent malaria. To make the experience tolerable, they mixed the tonic water with sugar and a slug of gin, and one of the most popular summer cocktails was born.

In the first half of the last century, several drugs that were functionally similar to quinine, including chloroquine, were discovered, and these became the materials of choice to treat and prevent malaria. However, these drugs represent an intense selection pressure on the malaria parasites, and consequently, resistance to chloroquine developed in the protozoans. Chloroquine resistance is now globally widespread. Fortunately, a new class of anti-malarial drugs, the artemisins, was developed in the 1970s. Artemisins are found in a Eurasian relative of the big-leaf sagebrush so common in the American West. Chinese herbalists had used the plant to treat malaria and other fevers for thousands of years. Unfortunately, resistance to this new class of medicines has also now been detected in Southeast Asia, and the race is on to find new ways to combat the disease before this resistance, too, becomes widespread. Resistance to malaria drugs almost always first appears in areas where the medications are used improperly or where they are poorly manufactured. Unless and until we come up with some new strategies for dealing with this devastating disease and its vectors, malaria is likely to continue to sicken and kill millions.

## The Black Death: Bubonic Plague

One of Edgar Allen Poe's most famous short stories is "The Mask of the Red Death," in which he describes the invasion of a sequestered group of nobles by a disease that kills all. Poe's disease is fictitious, but it was preceded by a real disease that wreaked havoc on European society for hundreds of years. This real disease was caused by a bacterium, *Yersina pestis*, and for most of its reign over Europe, it was simply called, "The Plague."

*Y. pestis* is thought to be native to the dry grasslands of western Asia, the steppes, where it originally was a disease of ground squirrels and gerbils, but it spread throughout Asia in antiquity as early trading routes were established. It moved because, since the earliest days of commerce, rats, specifically black or roof rats, have been transported along with the commodities of trade, and black rats almost always have Oriental rat fleas. Fleas are the vectors of plague, and the plague bacteria have an insidious effect on the flea that increases transmission. Once a flea bites an infected rodent, the bacteria proliferate in the "throat" of the flea, eventually completely occluding it and preventing the flea from ingesting the blood it needs to survive. The flea becomes frantically hungry, greatly increasing its bite frequency and also greatly increasing the probability that the bacterium is transmitted to another host, be it another rodent or a human. Eventually, the flea starves to death. While some rodent species appear to be somewhat more tolerant of the bacterium, many also die from the disease. This has an important ramification we'll address later. About half the people who contract plague die from it without modern medical treatment.

Plague has three forms of disease, all ghastly to experience, and all caused by the same bacterium. **Bubonic plague** is the most common form and the one that is transmitted by the bite of an infected flea. Bubonic plague is essentially an infection of the lymphatic system. The disease derives its name from a characteristic symptom, the **bubo** (Fig 17.5), which is a grossly inflamed lymph node, engorged with blood, plague bacteria and white blood cells. Other symptoms of bubonic plague are high fever, chills, and nausea. When the bacteria invade the bloodstream, the disease transforms to **septicemic plague**. Subcutaneous hemorrhages form as small blood vessels are destroyed and blood pools in the affected tissue. This blood coagulates, leaving the black lesions that give the disease its other name, "the black death." Depending on the course of infection and other factors, bubonic and septicemic plague can cause death in a matter of a few days. Should the bacteria invade the lungs, pneumonic plague is the result. This is the most deadly form of the disease and can readily be transmitted from one person to another through the air, as the sick person coughs and sneezes. Pneumonic plague can kill in as little as two days and usually within six days; untreated, it is invariably fatal.

Plague can be readily treated with modern antibiotics, and if the disease is diagnosed and treated promptly, most modern victims survive. However, we've only known what caused this disease since the late 1800s; prior to this, the plague was essentially untreatable and terrifying. There have been at least three major, historical outbreaks of plague. When such an outbreak spreads over a very large geographical area, it's called a **pandemic**.

**Figure 17.5.** A historic image depicting plague patients with bubos— swellings containing plague bacteria and white blood cells— and subcutaneous lesions.

The first of these major pandemics, The Plague of Justinian, occurred in the Middle East and around the Mediterranean, during 500–600 AD, and may have decreased the region's population by 50%. The plague then disappeared for almost a thousand years before re-emerging in the Middle Ages to spread across virtually all of Europe over the following 200 or so years. This was the **Great Plague** or Second Pandemic and again, over time, almost half of Europe's citizenry perished to the disease. The Great Plague had profound impacts on European society. Perhaps one of the most infamous ramifications was the rampant persecution of "the others," the Jews and Romany that were significant and conspicuous minorities throughout Europe. In many places, after the plague ran its course, there were no more Jews, not because of disease, but because of pogroms to eliminate them as the putative cause of this horrible disease.

By any measure, life during the Middle Ages was difficult and dangerous, but the constant threat of a calamitous wave of inexplicable, ghastly disease had a profound effect on people's outlook and aspirations. Prior to the emergence of the Great Plague, Europe's economies were based on a feudal system, in which a tiny minority of very wealthy people (the nobles) owned everything, and most of the population consisted of peasants who essentially share-cropped the nobles' lands. After the tremendous population declines caused by the waves of disease, however, the labor of the survivors became much more valuable, and the feudal system collapsed. Survivors also inherited the belongings of their late relatives, so they became inherently wealthier and began to demand greater educational opportunities. Since the disease struck all classes and occupations (though not necessarily equally), a large number of the Catholic priests in the region also died, and this coupled with some of the practices of the church, changed the populace's relationship with the church and may have contributed to the Protestant Reformation initiated by Martin Luther.

At the height of the plagues, in major cities, hundreds of people a day were succumbing to the disease, and while no one knew what caused the disease, they clearly understood it was contagious. Dealing with the literal mounds of dead bodies became an enormous task, and tens of thousands of victims ended up being unceremoniously and anonymously interred in mass graves, which themselves often went unmarked. Mass burials from the times of the plagues are frequently found today during construction projects, and they can provide valuable, if melancholy, insight into the epidemics.

Dealing with the living victims was also problematic, and this lead to the emergence of the plague doctor (Fig. 17.6). These largely untrained practitioners donned a supposedly protective outfit, consisting of a long leather frock, long leather gloves, a mask constructed so that one breathed through a "posey" of protective herbs, a hat and goggles, to examine the ill and provide what care they could. They typically touched their patients only with a long rod, and many of the treatments they administered like lancing the bubos probably did more harm than good. In spite of their protective outfits, being a plague doctor was a very risky, though lucrative, proposition, and most died themselves from the disease.

Other measures populations took included killing dogs and cats (as supposed carriers of the disease) and quarantining those who were infected by locking them, and often their families, inside their dwellings. Of course, both of these measures are largely ineffective and may have made things worse, because killing cats and dogs probably increased rat and flea populations. (It must be mentioned that cats can be short-term reservoirs for the disease.)

**Figure 17.6.** A plague doctor's wardrobe.

When the Great Plague subsided, Europe was decidedly different, but the disease wasn't done with the continent. It re-emerged in the 1600s and 1700s, as well. During a single week in September in London in 1665, over 7,000 people perished from the plague. Almost 70,000 people died of the disease that year, when London's population was about 450,000 people.

The Third Pandemic began in the 1800s, and this outbreak was indeed global, with large numbers of cases in Asia, Australia, Africa, parts of Europe, and parts of North and South America, where plague previously was unknown. Tens of millions of people died during this event, which did not officially end until 1959. However, it was during the Third Pandemic that the scientific community finally determined what caused it, which animals served as reservoirs and vectors, and how to treat infected persons. A vaccine for plague was developed during this time, and hundreds of thousands of people were vaccinated.

**Figure 17.7.** Black-footed ferret. Plague threatens other species besides humans and rodents.

Plague is not gone, although deaths to the disease have become extremely rare in most parts of the world. The disease remains endemic in western North America, where it persists in populations of ground squirrels, chipmunks, and prairie dogs. A handful of people are infected each year in the United States. Most of these folks are required to work around ground squirrels or their habitat, and most are treated rapidly and successfully. Plague in the western United States also poses a significant threat to at least one endangered species. The black-footed ferret (Fig. 17.7) feeds almost exclusively on prairie dogs and was driven almost to extinction by the persecution of these rodents as rangeland pests. Prairie dogs are highly susceptible to plague, and outbreaks of **sylvatic plague** (plague in squirrels) have wiped out prairie dog colonies where attempts to reintroduce ferrets were being conducted.

# The War-time Disease: Epidemic Typhus

It's a sad fact of human nature that for as long as our species has been around, tribes have settled disputes with other tribes through war. We have, over the last 100,000 years or so, created increasingly deadly ways to kill our enemies, transitioning from pointed sticks and rocks to drones and nuclear weapons. But for much of our history as a warring species, diseases, many transmitted by insects, have often had as much to do with who won a particular conflict as military strategy or sophisticated weaponry. Prominent among these maladies is epidemic typhus.

Epidemic typhus (not to be confused with typhoid fever and many other "typhus" diseases) is caused by a bacterium, *Rickettsia prowazekii*, closely related to the agent, *R. rickettsii* that causes Rocky Mountain spotted fever. The scientific name honors Henry Ricketts and Stanislaus von Prowazekii, two scientists who died, on separate continents, trying to identify the agent that caused this disease.

Epidemic typhus is transmitted by the human body louse. The bacteria proliferate in the insect's gut, and they defecate contaminated feces while they feed. Since lice bites are itchy, the victim scratches the bite, abrading the bacteria into the skin. The symptoms of epidemic typhus are similar to those of Rocky Mountain spotted fever: high fever, headache, and muscle pain, along with others. However, the characteristic, pin-point rash associated with epidemic typhus starts on the victim's trunk and spreads to the limbs, while the reverse occurs with Rocky Mountain spotted fever. Untreated, mortality can exceed 50%. As perhaps a small consolation, the louse typically dies from its infection, too.

The bacterium that causes the disease can easily be treated with modern antibiotics, and a successful vaccine was developed during World War II, but prior to these developments, there was little that could be done for those ill with disease. Most people who recover from it have life-long immunity, but in some folks, it goes into hiding, only reappearing, sometimes decades later, when such people experience illness or stress. These latent carriers often initiated outbreaks in the past, when times became difficult- as during times of conflict.

During war, combatants, resident civilians, and refugees are often forced to live in crowded conditions with little access to regular hygiene—conditions perfect for the proliferation of body lice. Should any of the people living in these crowded and stressful conditions have a latent typhus infection, the disease can enter the louse population, and be transmitted person to person by the lice they carry. Once such an outbreak gets started, it can sweep rapidly through a human population. This very thing has happened time after time throughout history.

The relationship between typhus and lice is apparently not that old, perhaps no more than 2,500 years, and so ancient records of typhus during times of war are sketchy and unreliable. Given the state of hygiene during the Middle Ages in Europe, no doubt typhus (along with a host of other diseases promoted by filth) was a factor during the many wars fought during that time, but it is difficult to establish its relative importance from the thin historical data. However, by the 15th century, a clear relationship between typhus and war was established. About the same time that Columbus was preparing to sail across the Atlantic, the Spanish were trying to drive the Moors from their country at Grenada. Typhus struck the besieging Spanish army of about 25,000 soldiers and felled 17,000 of them. As the survivors retreated, they spread typhus throughout Western Europe. Five times as many soldiers succumbed to the disease than were killed in battle! In Naples, Italy in 1525, a relatively small number of Spanish forces, themselves decimated by typhus, were saved from annihilation at the hands of the besieging French when the disease wiped out 25,000 French troops.

Thirty years later, the Holy Roman Emperor Maximillian II tried to wrest Hungary from the grip of the Ottoman Turks, but was crushed when his force of 80,000 troops was visited by typhus. One major battle during the Thirty Years War, in 1632 didn't actually happen because of typhus, when the Swedish and German forces were both made too sick to fight. Typhus continued to accompany every conflict in Europe. It also became a common and a devastating fact of life in the jails and prisons of the time (where it was called "gaol fever").

Perhaps the most notorious example of the impact of typhus on the course of warfare occurred in 1812, when Napoleon Bonaparte sought to conquer Russia with his Grande Armee of 600,000 men, after dominating most of the rest of Europe. Napoleon's army left Poland in June of that year. By the time they arrived in Moscow in September, his forces had been reduced to about 100,000. Even though he successfully "captured" Moscow, he couldn't hold it, and he was forced to retreat back to Poland. When his forces finally left Russian turf in December, only about 10,000 survived. Only 100,000 of the total lost had fallen in combat. The rest were conquered by typhus, dysentery, starvation, and exposure. Although Napoleon would rebuild his army and experience some additional victories, his military efforts never fully recovered from his typhus-assisted defeat in Russia, and Waterloo (where typhus again had an impact) sealed his fate.

Typhus continued its deadly role in conflicts through the remainder of the 19th century. During the Crimean War typhus and other diseases sickened six times as many soldiers as were killed by the conflict, and typhus was a gnawing presence in all the subsequent European wars of the 19th century. With the turn of the 20th century, however, things were about to change. By the onset of World War I, we had come to understand that body lice transmitted the disease, and programs were put in place by combatants on both sides to reduce lice populations and the potential for disease. These delousing procedures were time-consuming, however, and, during one phase of the war, may have delayed German troop movements to an extent that caused them ultimately to lose the war. In other words, modern Europe might look much different had it not been for typhus. Before World War I ended, the Tsar of Russia was overthrown and the resulting revolution, with all the turmoil, disruption, and displacement it entailed, created conditions perfect for lice and typhus, with perhaps three million Russians succumbing to the disease.

Typhus reemerged during World War II, but by this time the Allies had an effective vaccine and effective insecticides (notably DDT) to control the vectoring lice. During the war, for the first time in history, Allied forces were able to halt epidemics. However, typhus still claimed perhaps millions of lives, particularly in Nazi concentration camps, but also among the German civilian population.

Since World War II, typhus has been a much smaller concern in most conflicts, because of the availability of effective lousicides and effective medicines, However, as recently as 1997, an outbreak of typhus sickened 100,000 people and killed 15,000 during the civil war in the African state of Burundi. Typhus and other diseases are likely to always be at least a part of the misery inflicted by war.

We've only discussed three diseases vectored by insects, but its important to recognize that there are many, many others that have been historically significant (Table 17.1). No part of the inhabited world is without arthropod vectored diseases, and there is a great likelihood that many more will infiltrate human populations in the future, as our populations increase, we exploit more of the world's surface, and we move ever more efficiently and rapidly about the globe. We evolved in a context of insects and insect-vectored diseases, but our penchant for invention will ultimately result in our having to confront new diseases, transmitted by novel vectors.

**Table 17.1** *Some Additional Arthropod Vectored Diseases (there are many more. . .)*

| Disease | Vector |
| --- | --- |
| Tapeworms (some species) | Fleas |
| Filariasis | Mosquitoes |
| Onchocerciasis (river blindness) | Blackflies |
| Trypanosomiasis (sleeping sickness) | Tsetse flies |
| Chagas Disease | Triatomine assassin ("kissing") bugs |
| Yellow fever | *Aedes* mosquitoes |
| Chickangunya | Mosquitoes |
| Zika | *Aedes* Mosquitoes |
| Dengue ("break-bone fever") | *Aedes* Mosquitoes |
| West Nile Encephalitis | Mosquitoes |
| Scrub Typhus | Chigger mites |
| *Leishmaniasis* | Sand flies |
| Relapsing fever | Ticks, human body louse |

# XVIII

# They'll Even Eat Fried Green Tomatoes!
## Insects and Agriculture

In order to live, we must eat, and evolution has dictated that we are omnivores who require both plant-based and animal-based foods. For most of our history as a species, we acquired the foods we needed by harvesting them from the environment around us. We were hunter-gatherers, relying on the animals we could kill for protein and the plant products we could forage for carbohydrates and other important nutrients. A hunting and gathering lifestyle is sustainable only if the human group engaged in it is relatively small. A given area of land (and water) can only support so many humans if they are foraging for their nutrients, and no more.

About 10,000 years ago, however, groups of people in several areas of the world developed a new way to produce food—they invented agriculture. Agriculture is the deliberate manipulation of plants and animals to increase food production. This development was not simultaneous in all areas. World-wide, agricultural production was adopted over a period of about 5,000 years, and many groups of people in various places maintained hunting and gathering lifestyles well into the 20th century. The big advantage provided by agriculture is that it allows much larger human populations

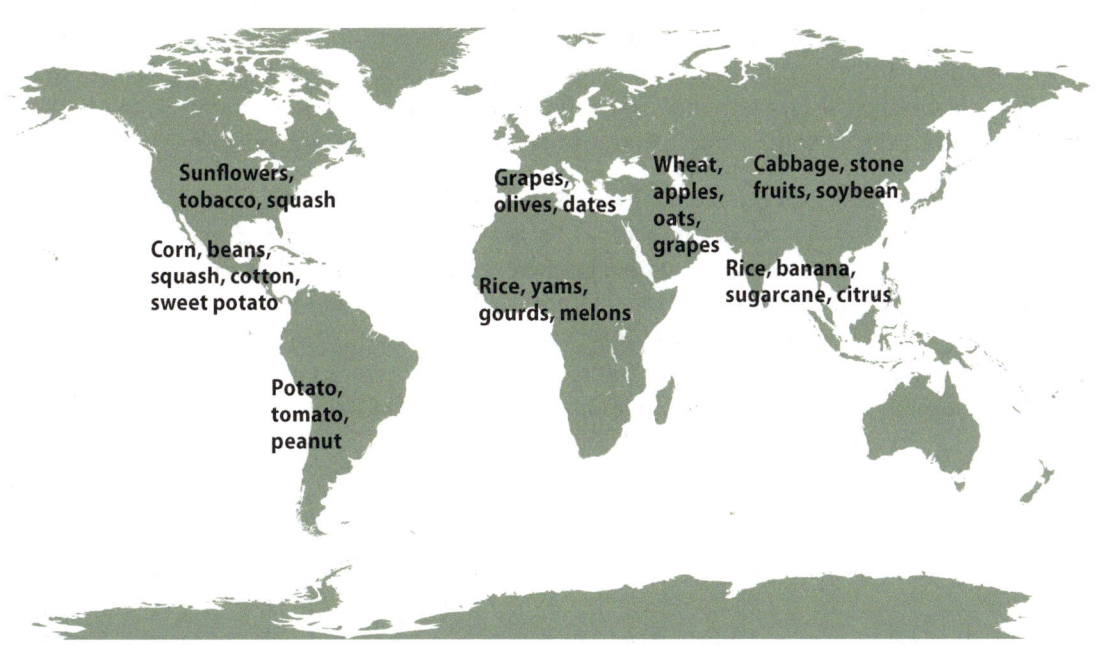

**Figure 18.1.** World map showing centers of domestication for some major plant crops.

to occupy a given area of land as long as that land is productive. There are a large number of often conflicting theories about exactly what drove the development of agriculture which we don't have the space to get into here, although climate change may well have been an important driver. Virtually all the crops we eat on a day-to-day basis were domesticated in about 10 areas of the world (Fig. 18.1). Note that most of what we eat comes from places outside North America.

For most of our history as agriculturalists, approximately 9,800 of those 10,000 or so years, we were **subsistence farmers**, basically growing what we and our families needed to survive, and very little more. Many people in developing parts of the world remain subsistence farmers. In subsistence agriculture, one or two staple crops provide most of the calories to the diet, but smaller quantities of a great many other crops may be grown as well. Often, animals are also raised for labor and for protein. The size of individual farm units tends to be relatively small, on the scale of tens of acres of land per farm. The family that lives on the farm provides almost all of the labor that goes into production on that farm. A landscape that sustains a population of subsistence farmers, therefore, is often a very heterogeneous landscape with a diversity of crops in a wide range of developmental stages. This has significant impacts on pest insects in such a landscape, and we'll discuss these later.

About 250 years ago, one of the greatest transitions in the history of humanity began with the emergence of the Industrial Revolution, first in England and then, over the following centuries, throughout much of the rest of the world. The development of technologies for manufacturing goods drove demand for labor, and people began leaving the rural areas where they engaged in subsistence farming for cities, where employment, and perhaps a more secure future, could be found. In order to feed these growing masses of people, a concurrent and similarly earth-shaking transition began in agriculture. The shrinking numbers of people left on the farm had to figure out how to grow food not only for themselves, but also enough to feed those moving to the cities. The answer was agricultural technology—the development of tools and tactics that enabled the intensification of agriculture—increased yields and allowed still more people to abandon the farm for the city in a positive feedback loop. We have transitioned, in most of the developed world and over the last couple hundred years, from societies composed of rural populations engaged in subsistence agriculture, to societies composed of large, concentrated industrial populations fed by a tiny proportion of the population engaged in **commercial agriculture**.

In the United States, in 1790, 94% of the population lived in rural areas and was engaged in agriculture in one way or another. Today, less than 2% of the population is actively engaged in growing food and fiber. Today's commercial agriculture looks much different from the subsistence systems we used for those many millennia. In today's systems, crops are typically grown as single-species **monocultures** in much larger fields, with most of the inputs either mechanical or chemical, and the livestock we raise for protein are concentrated in large production facilities reliant on grain production for feed.

While the vast majority of the 1,000,000 or so species of insects are beneficial, in that they cause us no harm and sustain ecosystems, insects, and other arthropods, have been competitors with us for the food we grow since the beginnings of agriculture. (In most places, before the advent of farming, insects were most likely regarded mostly as another food source, but more on that later.) Every crop we grow has a large number of arthropods that have evolved to exploit it as a host. A crop field, for those insects that are adapted to that plant species, must represent something close to heaven—virtually unlimited resources that are very easy to find. Under the right environmental conditions, the populations of these insects can explode, with potentially disastrous consequence for the farmer. In the United States, there are about 200 or so species of insects that are regular, significant pests of one or more crops, and a few hundred others that are occasional pests.

As I said earlier, most of the crops that we eat from day to day originated in another part of the world other than North America and are therefore exotic to this continent. Even most of the crops that Native Americans grew before European Colonization came from Central America. Most of these crops reached our shores hundreds of years ago, when transport was much more arduous, and when awareness of the hazards of introducing exotic organisms to new landscapes was essentially nonexistent. Consequently, over the last few hundred years, we have accidently introduced many, many species of insects (and other organisms) that have subsequently become important pests. Often, these species become much more severe pests here than in their countries of origin, because, while they are accidently introduced, the beneficial organisms that regulate their populations back home are not. About 40% of our significant pests are exotic to North America, and, with our greatly enhanced, modern abilities to transport stuff all over the globe, we are constantly introducing even more.

# Pests of Crop Plants

Insects consume about one third of the crops we grow globally, in spite of all that we do to prevent this. Approximately half of all insect species are phytophagous or plant eaters. Of course, the vast majority of these are not economically significant to us because they eat plants that we don't use. Phytophagous insects come primarily from nine of the 30 or so orders. This imbalance is due to the fact that three of the four megadiverse orders (Coleoptera, Lepidoptera, and Diptera) all contain large numbers of plant feeders. About half of the Hemiptera and virtually all of the Orthoptera are also plant feeders. Insects can damage plants and cause yield loss a number of different ways.

**Figure 18.2.** Grasshopper eating foliage.

Most crop pests, but certainly not all, have chewing mouthparts and cause their damage by physically destroying plant tissue. The Coleoptera (beetles) and Lepidoptera (moths and butterflies) are probably the two most important crop pest orders because of this. In the beetles, both the adults and the larvae typically have chewing mouthparts and therefore are potentially pestiferous. In the Lepidoptera, only the larvae, or caterpillars, have chewing mouthparts and can consume plant tissues. Insects with chewing mouthparts can damage plants by consuming the foliage (Fig. 18.2), eating the roots (Fig.18.3), eating the fruit (Fig. 18.4), tunneling into stems or other plant parts (Fig. 18.5), or even mining the green tissue from the middle of the leaves (Fig. 18.6). The relative importance of a chewing pest is often dependent on the plant part they attack; in most cases we are most concerned about the "direct pests" that feed on the plant part we are interested in harvesting. The "indirect pests" that feed on other parts of the plant are often, but not always, a smaller concern. In an apple orchard, we are most concerned about those insects that feed on apples and less worried about those that feed on leaves or roots.

**Figure 18.3.** A sweet potato damaged wireworm larvae feeding.

A smaller, but still important, percentage of insects feed on plant with piercing-sucking mouthparts. Most of these insects are in the order Hemiptera, and they

**Figure 18.4.** Green June beetles devouring a peach.

They'll Even Eat Fried Green Tomatoes! Insects and Agriculture   157 fall into two major groups: the "true bugs," distinguished by their hemelytral forewings (Fig. 18.7), and the aphids and their allies (Fig. 18.8). Most of the pest species of the phytophagous "true bugs" are problems because they are direct feeders on the fruit of the crop plants. Aphids and their allies often feed primarily on the leaves and stems of the plant, but, because of their tremendous reproductive potential and their ability to transmit potentially devastating plant diseases, they are the most important pests in some cropping systems.

*The economic injury concept.* Insect pest populations are extremely dynamic. A particular species may languish at very low levels for weeks, months, or even years, and then erupt to damaging levels when environmental conditions are right. Others occur in damaging levels every year at least in some places, but those places may vary. Because insect populations are so changeable, it is critically important for those engaged in agriculture to be systematically vigilant for the important pests in each crop they grow. We call the system of crop pest surveillance "pest scouting," and we use the information gleaned from this regular, systematic, and rigorous sampling to guide our response to the pests. Every pest species can occur in a crop field at levels so low that we can't detect any measurable effect of the insects on the yield of the crop. Conversely, they can also reach populations that can completely destroy the utility of the crop. Between those two extremes, the impact of the pest varies with the intensity of the pest population (Fig. 18.9). The most important population level for agricultural producers to know is the population level that causes damage to the crop *equal* in value to the cost of controlling the pest through some immediate action, like spraying an insecticide on the crop. Any population below this level is going to cause less damage than can be recovered through the control operation, and we'll regard it as a *sub-economic* population. Any population above this level is causing economic loss we can

**Figure 18.5.** Squash-vine borer tunneling in squash stem.

**Figure 18.6.** Tobacco splitworm caterpillars mining the lamina in a tobacco leaf.

**Figure 18.7.** A green stinkbug.

**Figure 18.8.** Aphids on a leaf petiole.

economically recover, and we'll act against that population. This population level is called the Economic Injury Level (EIL). In actual practice, an action threshold is established below the EIL to account for the potential population increase in the pest that may occur between detecting the damaging population and actually doing something about it. Farmers use this concept to avoid wasting time, money, and resources on unnecessary control applications. They only treat a crop when they can recover economically significant amounts of yield. This also greatly reduces the environmental impact of agriculture by reducing the amounts of things like pesticides that might be introduced into the ecosystem.

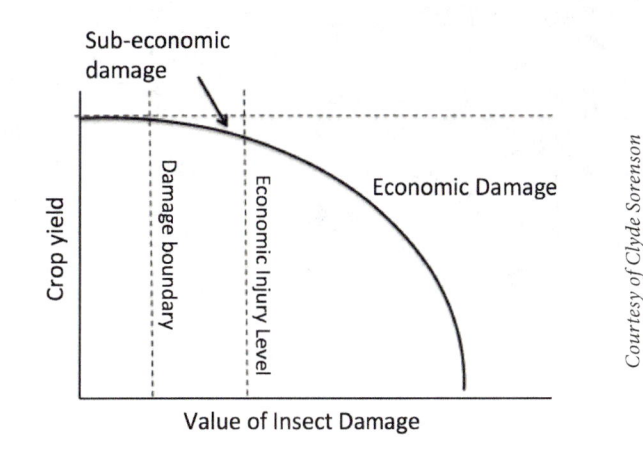

**Figure 18.9.** The relationship between insect damage and crop yield and quality.

## Pests of Livestock

"So nat'ralists observe, a flea
Has smaller fleas that on him prey;
And these have smaller fleas to bite 'em.
And so proceeds *Ad infinitum*." Jonathan Swift

Livestock, like all animals, including us, are bedeviled by a host of insects and other arthropods, and many of the same insects that bother us, bother our domestic animals. Probably the most important order of livestock pests is the Diptera—the same true flies that so aggravate us. Mosquitoes, blackflies, horseflies, deerflies, and other biting flies can pose significant problems to livestock producers by reducing how efficiently the animals feed, and, thus, the pounds of meat they put on, gallons of milk they produce, or number of eggs they lay. Particular species of flies, however, can cause even more grievous injury. Myiasis is the condition in which fly larvae, or maggots, inhabit living vertebrate tissue, and many fly species cause economically significant myiasis in livestock species. Below, we'll describe three prominent examples.

*The primary screwworm.* The primary screwworm is an unexceptional-looking fly (Fig. 18.10) that can cause extraordinary harm to vertebrate animals. Female screwworms lay their eggs in their hundreds on open wounds on living animals. Wounds as minor as a scratch from a thorn or barbed wire may attract them, and the fresh umbilical wound on a new-borne calf or lamb is particularly attractive. The eggs hatch and the maggots feed on the damaged tissue in the wound and the healthy tissue surrounding it, gradually enlarging the wound and potentially attracting additional flies. Eventually, the wound may grow so large that the animal succumbs (Fig. 18.11). At one time, screwworms caused millions of dollars of economic loss to livestock producers in the southern United States, but no longer. The screwworm is not a problem for modern southern ranchers because the insect has been eradicated from the United States and Central America. How this was done is an extremely interesting story.

The tale begins in the 1930s, when two USDA entomologists, E. F. Knipling and R. C. Bushland, theorized that a pest insect species could be driven to extinction if a large enough number of sterile males could be released into the population. The idea, called the Sterile Insect Technique (SIT), was that wild, fertile females that mated with the sterile males would produce

**Figure 18.10.** Adult Primary Screwworms, *Cochliomyia hominovorax*.

**Figure 18.11.** Screwworm lesion.

**Figure 18.12.** An adult specimen of the northern cattle grub.

no offspring, and, if the releases were continued over enough generations, eventually the few remaining females would have only sterile males to mate with, resulting in the collapse of the population. As it turns out, the screwworm was the perfect candidate for this theory, since it occurred in the landscape at relatively low native populations (in spite of its tremendous economic harm) and the females only mate once. Just as importantly, after World War II, they developed a way to produce sterile males by exposing them to radiation as pupae. The resulting males were in every other way normal in their behavior and longevity. In the early 1960s, the USDA released several billion sterile males across Florida, south Georgia, and Alabama, and completely eradicated the screwworm from this immense area in 18 months. Over the ensuing decades, the program was continued until the insect was eradicated from all of North and Central America north of Panama. Releases continue to this day in Panama to prevent the reinvasion of the north, and Knipling and Bushland received the World Food Prize in 1992 for the food security their invention produced. SIT has since been used to control populations of some other pests in other parts of the world. Many are working on ways to expand this technology to other important medical and veterinary pests through new genetic engineering approaches.

*The northern cattle grub.* Cattle grub flies are large, somewhat bee-like flies (Fig. 18.12) that lay their eggs on the legs of cattle. Some folks don't give cattle a lot of intellectual credit, but cattle do seem to realize that the big flies buzzing around their legs are bad news, and, rather than eating and putting on weight as they should, they spend a lot of time running and kicking (a behavior called "gadding") to avoid the harassing insects. The eggs hatch in about four days, and the tiny larvae burrow into the skin. As they grow and develop, they migrate through the animal's connective tissues below the skin, leaving paths of destruction and scar tissue in their wakes; first to the exterior of the esophagus and spinal column, and ultimately to the skin of the cow's back, where they cut a breathing hole (Fig. 18.13) and finish their larval life stage. Mature larvae crawl through the breathing hole, drop to the ground, and pupate, to start the cycle anew next year. Cattle grubs cause at least three kinds of loss for cattle producers. First, the cattle put on less weight when they are gadding; second, the hide is ruined for leather, since the scars from the breathing holes can't be removed; and third and most important, their migration leaves tissue damage and scar tissue that substantially reduces the value of affected meat. Cattle grubs can be controlled with appropriate medicinal insecticides, but the applications have to be timed properly, and control can be incomplete.

*Horse bots.* The three species of horse bots are also largish flies that resemble bees (Fig. 18.14). Depending on the species, they lay their eggs on the legs, chest, or muzzle of the horse. Like cattle grub flies, horse bot adults startle horses and cause them to gad, potentially injuring themselves or nearby humans. The eggs hatch in response to the warmth and moisture of the horse's tongue as it grooms itself, and the tiny larvae lodge in the animal's gums for several weeks before migrating to the stomach where they complete their development. Large numbers of horse bot larvae in a horse's stomach can cause serious problems; the larvae themselves are stealing blood and nutrients from the animal's stomach lining, and the insects' sharp mouth hooks irritate and damage the tissue. They can also obstruct the digestive tract, causing potentially fatal colic. Once the larvae finish their larval development, they hitch a ride on the poop train and end up deposited with the dung on the ground, where they burrow into the soil and pupate. As with cattle grubs, judicious use of appropriate medicinal insecticides can control these creatures.

Apart from flies, other arthropods can be significant pests of livestock. Chewing and sucking lice are very serious problems in poultry, swine, and cattle, particularly in confined livestock operations and during the winter. Various species of ticks can severely compromise the health of cattle, both through direct blood loss and through their roles as vectors of important livestock diseases.

**Figure 18.13.** a. A cattle carcass showing large numbers of cattle grubs under the skin; b. Cattle grub breathing holes left in a cowhide (Both images courtesy Agriculture and Agri-Food Canada).

**Figure 18.14.** An adult specimen of one of the Horse Bot flies.

Many serious livestock diseases are transmitted by arthropods. When I worked in Nevada, we conducted studies of a large soft tick called the pajarello, which transmitted a bacterial cattle disease called Foothill Abortion. We also conducted surveillance for the tiny biting midges that transmit the potentially devastating viral disease called bluetongue to sheep. Mosquitoes vector numerous viral diseases to horses and other livestock, including equine encephalitis diseases that are often fatal.

# Pests of Our Household Livestock: Arthropods that Attack Our Pets

**Figure 18.15.** Pamlico and Hatteras, my chocolate Labrador retrievers.

**Figure 18.16.** Cat flea.

**Figure 18.17.** Brown dog tick.

We've always had pets in our house, as I'm sure is true for many of you. In my house, some of the most cherished occupants are our three cats (Eno—after the river; Catawba—Cat, for short; and Eli—just Eli) and two chocolate labs, Pamlico and Hatteras (Fig. 18.15). Our furry friends, like all animals, are potential hosts to arthropod parasites.

Some of the most bedeviling pests of pets are fleas. While there are many species of fleas that might bite dogs and cats, by far the most common pest flea in North America is the cat flea (Fig. 18.16). Like most other fleas, only the adults (both sexes) bite. The larvae live in the bedding of the host animal and feed primarily on the feces of the adult fleas. Cat fleas have very catholic tastes and are quite likely to bite humans as well as the dogs and cats that live with us. Over the last 20 or so years, we have developed some very effective flea management programs for our pets and our homes, based on topical insecticides we apply to the animals and including other products, and, for many of us, fleas are currently a minor concern. However, there is evidence that populations of the cat flea are acquiring resistance to some of the highly effective insecticides we currently use, and they may again become significant nuisances in the future.

Ticks are also significant pests of pets, and ticks transported by pets can pose a threat to pet owners. Pets may be attacked by all three of the tick species we discussed previously. Additionally, dogs also may be bitten by the brown dog tick (Fig. 18.17). All three life stages of the brown dog tick prefer dogs as hosts, and, because of this, they can complete their entire lifecycle inside our dwellings, unlike the other tick species, so long as a dog remains a resident of the house. Ticks can also transmit disease to pets, just as they do to humans. Dogs can contract Lyme disease, Rocky Mountain spotted fever, and erlichiosis from tick bites. Many of the same materials that effectively control fleas on pets are effective against ticks as well.

Perhaps the gravest threat posed by an insect to our pets is the dog heartworm transmitted by various species of mosquitoes. The agent in heartworm is a large roundworm, *Diroflilaria immitis*. Infective larvae exit the mosquito when it bites an animal and enter the skin through the bite site. Over the course of about six months, the larvae grow and slowly migrate, through the dog's vascular system, to the animal's heart, where they complete their development and mate. Females then release microfilaria, which circulate in the dog's vascular system until another mosquito picks them up during a blood meal. The microfilaria transform into infective larvae in the mosquito, and the cycle starts over. The adult worms can live for years in the heart, producing new microfilaria, and, over time, their presence damages the heart muscle, lungs, and kidneys. Infected dogs invariably die before their time. Diseased dogs are lethargic, have a chronic cough, and have low endurance; their tongues and mouth lining may become bluish after exercise due to poor blood flow. Heartworm is most common where mosquitoes are most common, but they are found throughout North America. While adult worms can be cleared from infected dogs with a medication containing arsenic, this is a risky procedure that can result in permanent injury or even death. The best way to protect your dog from this peril is to have it tested for heartworm and then keep it, year round, on a regimen of preventative medication.

Other animals can also be infected with heartworms. Wild canines (and free-roaming or untreated dogs) are probably the most important reservoir for the agent, but raccoons, mustelids (weasels and their kin), and cats can also become infected. Cats are not a good host, though, and many are able to clear infections spontaneously thanks to their immune systems. On rare occasions, humans may be infected, but the worms can't complete their development in humans and typically die before they cause illness.

# XIX

# There's a Fly in My Soup: Stored Product and Urban Pests

Ever since we stepped into that first cave a human called home, we've been looking for ways to make more comfortable and secure dwellings. Over that time, as our houses and other buildings became sturdier, and the environment inside them became more stable, we've created an interior environment that is not only ideal for us, but one that is perfect for large numbers of other creatures. No one in the world lives alone; we all have large numbers of tiny houseguests living with us. Our homes, storage facilities, and businesses offer a uniform temperature year round, ample water, and usually ample food. Most of the creatures exploiting these resources are completely harmless, but some do end up being pests, for one reason or another.

## Pests of Stored Products

Insects have evolved to exploit virtually all organic resources in terrestrial and freshwater aquatic ecosystems. If something is alive, was once alive, or was made by something alive, there is probably an insect species, or perhaps many species, that exploit it. As we discussed in the previous chapter, there are large numbers of species of insects that exploit growing crops, but there are also many insects that exploit the harvest, as well. Stored grains, like corn, wheat, rice, and soybeans, all can support diverse communities of seed-feeding insects.

Perhaps the most significant stored grain pests are beetles (Fig. 19.1). There are several dozen species of grain beetles, mealworms, and weevils that attack these products, each with its own specific preferences for grain type and condition. Because the environment in a grain bin or tank is so protected and consistent, grain beetles can demonstrate phenomenal, explosive population growth. Over the course of just six months, a single mated female of some species can produce a population of near 50 billion offspring! With most of these pest beetle species, both the larvae and the adults can damage grain. One of the most important consequences of infestation by insects, apart from the actual loss of grain consumed, is that the metabolic activity of large numbers of small creatures changes the

**Figure 19.1.** Rice weevil— one of many species of grain-infesting weevils and beetles.

environment in the grain mass. Humidity, generated by the respiration of the insects, increases, allowing fungi to flourish and contaminate even more grain, often with extremely dangerous mycotoxins. Sanitation around grain storage facilities is a critically important task to limit these impacts.

Several species of small moths also attack stored grain and grain products, in addition to grain-infesting beetles, however, with these insects, virtually all the damage is caused by the caterpillars. The Indian meal moth (Fig. 19.2) also frequently becomes a particularly annoying pantry pest in the home, where it can infest breakfast cereals, dried pasta, and other grain-based products.

Other stored goods are also susceptible to insect attack. Many of the same insects that attack grain also infest spices and nuts. At least half a dozen insects can exploit dried tobacco products. Certain beetles and flies even attack stored country hams and cheeses!

**Figure 19.2.** Indian meal moth—the adult stage of a grain-infesting species of caterpillar.

## Pests of Fiber Products

Most of the organic fiber we use to make cloth, paper, and other similar products comes from either plants (cellulose) or animals (keratin). Cellulose is a polymer made of repeating units of the sugar glucose and is the primary structural material plants use to support themselves. Cotton is essentially pure cellulose organized as fine fibers; linen is cellulose harvested from the stems of the flax plant. Either can be spun into yarn, and the yarn then used to make fabric. Both can also be used to make paper and the like. A great many insects have evolved to utilize cellulose as food, usually with the assistance of symbiotic organisms living in their guts. Most of them feed on wood in nature, but these same creatures are just as happy to eat cellulose fiber products if given the opportunity. We'll be discussing wood-destroying pests more in a bit. Some products made with cellulose fiber are vulnerable to additional pests. Silverfish and cockroaches, for instance, will devour the glues used to bind books, and will even eat the pages of old books that have been much handled.

**Figure 19.3.** A dermestid beetle.

Keratins are protein polymers that vertebrates use in their protective coverings. Keratin provides the toughness in skin, and many of the substances that "decorate" vertebrate skin are composed of keratin. Hair, fur, wool, feathers, fingernails and claws, rhino horns, cow horns, and some fish scales are all made of keratins. Keratin, as a polymer and being like cellulose, is very difficult to digest, but a number of insects, notably the dermesitid ("skin") beetles (Fig. 19.3) and clothes moths (Fig. 19.4), have evolved the digestive enzymes and symbiotic relationships that allow them to survive on a diet of this substance. In nature, these insects play a very valuable ecological role as the last of the "mortuary crew," the creatures that come in and clean up the last dried bits of hide, fur, or feathers that cling to the skeletal remains after other insects have consumed all the other tissue that once was a dead animal.

**Figure 19.4.** Clothes moth.

166    Insects and People

Essentially any product that contains keratin, as well as any other kind of desiccated animal tissue, is vulnerable to exploitation by the keratin-eaters. Woolen clothing and other keratin products in storage can be protected with mothballs or crystals. Particularly valuable items, like fur coats, can be stored in refrigerated lockers. These same insects are also general scavengers on any dried, dead animals, and they can be the bane of insect collection and natural history museum managers.

# Wood Destroying Insects

Wood, of course, is primarily cellulose, and, as we've already mentioned, a great many insects feed on cellulose. In nature, these insects play a critical, valuable role in recycling the nutrients tied up in dead plant bodies, but in the home, they can cause catastrophic damage. Far and away the most economically important wood-eating insects are termites. In the United States, the most common and important species are the various subterranean termites (Fig 19.5). Subterranean termites nest in the ground, usually near a wood source, in colonies of a couple hundred thousand, and they build mud tubes to access wood that is located above the soil (Fig. 19.6). An average termite colony might consume about 12 pounds of wood per year; while this doesn't sound like very much, an undetected colony, over the span of just a couple of years, can cause thousands of dollars of structural damage to a home if that 15 to 20 pounds of missing wood is strategically located. Collectively, termites are responsible for billions of dollars in damage, repair, and control costs each year. Termites are found pretty much throughout the United States, although they are uncommon in deserts and absent at high altitudes. Virtually all wooden structures in the country should receive regular termite inspection and control.

**Figure 19.5.** Subterranean termites.

**Figure 19.6.** Termite mud tube.

A large number of beetles also can infest wood products, and although they rarely do serious structural damage to buildings, they can cause very significant cosmetic damage to wood. Powderpost beetles often infest products made of hardwoods that haven't been dried properly. The adults leave tiny "shotholes" (Fig. 19.7) when they leave infested wood, and the larvae produce dry frass the consistency of flour. Old house borers (Fig. 19.8) infest softwood, usually rough-cut and unpainted timber, and can consume large quantities of wood if left unchecked. Deathwatch beetles (Fig. 19.9) attack old, often damp softwood timbers. They get their name from the rhythmic clicking sounds they make as they seek mates; during Victorian times, these sounds were regarded as omens of an impending death in the family, since they were often most noticeable when folks were keeping watch overnight with a recently deceased person.

**Figure 19.7.** Powder post beetle "shot holes" caused by exiting adults.

There's a Fly in My Soup: Stored Product and Urban Pests

**Figure 19.8.** Oldhouse borer.

**Figure 19.9.** Deathwatch beetle.

**Figure 19.10.** A carpenter ant.

**Figure 19.11.** A carpenter bee.

A few Hymenoptera also infest wood, but not as a food source. These insects excavate in wood as a nesting substrate. Carpenter ants (Fig. 19.10) frequently exploit damp or fungus-infected wood in structures. If left undetected, they can cause significant structural damage, but they should be regarded as a symptom of other, more serious problems that lead to the wet, soft wood they prefer. Female large carpenter bees (Fig. 19.11) excavate their "T"-shaped brood tunnels in horizontal surfaces on the undersides of unpainted timbers. A pile of sawdust under the round nest entrance often indicates their presence. Carpenter bees don't generally cause structural damage, although they can cause significant cosmetic damage. Many folks are also intimidated by the large bees patrolling back and forth in front of porches or other structures offering suitable habitat. However, the patrolling bees are males looking for mates and are completely harmless. Painting wood with a good quality exterior paint usually discourages carpenter bees from nesting.

# Teaming Hordes: Household Pest Ants

No doubt to the alarm of many of you, most houses harbor a large number of arthropods of a surprising diversity, but only a relative few do things that elevate them to pest status. Indeed, a number of the arthropods living in your home should be regarded as beneficial allies, since they feed on those we regard as pests. Probably the most important groups of pest insects in the house are ants and cockroaches.

Ants become pestiferous in houses for some of the same reasons they are so successful: they are adaptable, they feed on a great diversity of substances, and they can be incredibly abundant. While ant colonies can range in size from just a few dozen workers to hundreds of millions, the average ant colony probably has several tens of thousands of individuals. Ants often forage at relatively great distances from the colony, and once a few workers have identified a promising food source, their trail marking pheromones will recruit hundreds of others to it. Pest species range in size from the relatively large carpenter ants (ca. one fourth of an inch long) to minute odorous house ants and pharaoh ants only a millimeter or so long. Ants are problematic because they potentially contaminate food or clean surfaces (both with microbes they may carry and with their own little bodies), and create a significant nuisance just in terms of their presence. Additionally, some species sting and pose a medical threat.

Roughly half of the house pest ant species we deal with in North America are native species that have adapted to modern dwellings, while the other half are exotic species that were accidently introduced from other parts of the world because they were already associated with humans. Different species of ants are attracted to different nesting habitats and to different food sources. Some prefer protein-rich foods, while others prefer sugar-rich foods. Their preferences can vary within a species during different parts of the year, making control of nuisance ants difficult. An insecticide bait that works for one species may not work for another, or may not work during some parts of the year.

Ants and termites found in the home are frequently confused because the winged reproductives of both are somewhat similar in appearance. However, it is easy to distinguish them if one focuses on a couple important charcteristics (Fig 19.12). Winged ants have fore and hindwings that are dissimilar in shape, a narrow "wasp waist" between the thorax and the abdomen, and elbowed antennae, while termites have fore and hindwings roughly the same size and shape, a thick "waist," and straight, bead-chain antennae. The presence of winged termites in the house is a very grave concern, since it probably indicates the presence of an active termite infestation, while the presence of winged ants (unless they are carpenter ants), probably represents more of a nuisance.

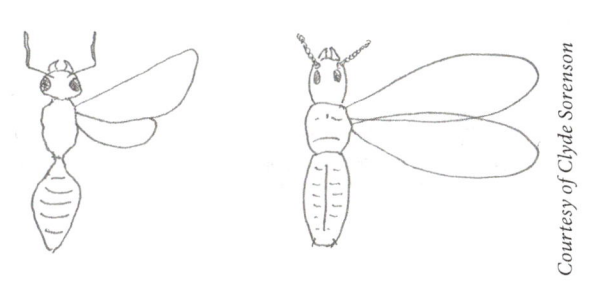

**Figure 19.12.** Diagram depicting differences between ants (L) and termites (R). Note that ants have narrow "waists," elbowed antennae, and forewings (if winged) much larger than the hind wings, while termites have thick "waists," straight, "bead-chain" antennae, and fore- and hind wings roughly the same size and shape.

## The Least Welcome House Guests: Cockroaches

There can be no doubt that the insects most abhorred in the house are the cockroaches. Almost everything about cockroaches makes them "icky." They are smelly, fast, "buggy"-looking creatures of the night that have long been associated with filth. However, it is important to reiterate that of the 5,000 or so species of cockroaches in the world, only about 20 or so are serious pests somewhere in the world. The rest should be regarded as beneficial, and sometimes beautiful, wildlife (Fig. 19.13). Of the pest species, four or five are common problems in North America; none of these are native to this continent.

**Figure 19.13.** A tropical, non-pestiferous cockroach.

**Figure 19.14** German cockroach.

**Figure 19.15.** American cockroach.

**Figure 19.16.** Smoky brown cockroach.

Far and away the most important pest cockroach in North America, and, for that matter, the world, is the German cockroach (Fig. 19.14). German cockroaches are the smallest of the common pest species. They are about a half-inch long, light brown, with two dark marks on the pronotal shield that protects their head. German roaches are not native to Germany, but probably originated in North Africa or Asia. This is a completely peridomestic species. They are not found in "the wild" anywhere in the world, and can only survive in the dwellings of humans and their livestock. Germans are the most common household roach, and in moderate infestations, are most likely to live in the kitchen where food and water are readily available. This species has the highest reproductive potential of the common pest species and can build to extraordinary populations numbering in the tens of thousands in individual apartments if conditions are right. Their small size allows them to stow away in grocery bags and other packaging, resulting in new infestations. German cockroaches contaminate food, transmit diseases to humans and pets, and are responsible for a significant proportion of cases of asthma in children. They are also extremely difficult to control and have acquired resistance to some of the insecticides used to control them. Some populations have acquired **behavioral resistance** to the toxic baits use to great effect against them. These roaches have evolved aversion to glucose, the sugar used to attract the roaches to the insecticide-laced baits.

The American cockroach (Fig. 19.15) is the largest common pest roach, at nearly two inches in length, and the one most typically associated with commercial establishments like restaurants, hotels, hospitals, and industrial food processing sites. American roaches are not native to the Americas but rather to Africa. The fact that they beat the European scientists classifying life to the Americas by a hundred or more years is a testament to their adaptability and ubiquity. Like the other pest roaches, American cockroaches produce a very disagreeable and highly recognizable odor, and like the others, they can foul food and other materials. Americans can fly well over moderate distances and are among the fastest running insects in the world. Since they are large, fairly primitive, and common, American cockroaches are also favorite research animals in biological and entomological studies.

Closely related to the American cockroach, and also native to Africa, is the smoky brown cockroach (Fig. 19.16). It is slightly smaller than the American and a uniform dark chocolate brown in color. Smoky browns prefer to remain outdoors and aren't nearly as pestiferous as Americans, although in very humid areas, such as southern beach communities, they can reach very high populations and become common in homes and outdoor living spaces.

The fourth common pest roach is the Oriental cockroach (Fig. 19.17). This species is slightly larger than the German, stockier in body shape, and almost black in color. Oriental cockroaches, oddly enough, are actually native to western Asia. These insects appear to need a very high humidity environment and so tend to gravitate to the dampest areas in the home, including bathrooms and cabinets under kitchen sinks. This tendency has encouraged many folks to call them "waterbugs" (perhaps to avoid acknowledging that they have a cockroach infestation in the home). Male Orientals have wings that almost completely cover the abdomen and they are capable of short flights, while the stubby wings of the females are non-functional. Like the other pest species, they produce a disagreeable smell.

There are other pest roaches in North America with more limited distributions, including the brown-banded cockroach and the Asian cockroach, which is virtually indistinguishable from the German roach. It's a virtual certainty that other pest species will find their way to North America over the coming years, as we continue to improve our ability to move commodities around the globe.

**Figure 19.17.** Oriental cockroach.

## "Killer Bees": the Africanized Honey bee

One final urban pest we ought to discuss, if for no other reason than the sensationalism surrounding it, is the Africanized honey bee. The honey bee, *Apis mellifera*, is not native to North America. The populations we are familiar with, and which we rely on for pollination services in many of our crops, originated in Europe and western Asia. However, the honey bee is native to essentially all of the Old World—Europe, Asia, and Africa, and dozens of subspecies have evolved to adapt to the huge range of climates and environments presented by this huge landmass. The subspecies native to Africa are more tolerant of the high temperatures and high humidities found in tropical and subtropical regions than the European and Asian subspecies that evolved in more temperate climes. In the 1950s, apicultural entomologists in Brazil were attempting to breed honey bees that would be better suited to the hot, humid climate of northern South America by making experimental crosses between the European honey bees common in apiculture and African honey bees adapted to the climate. An assistant accidently released some of these Africanized bees, and in the ensuing decades, they have spread throughout South and Central America and into the southern United States. Matings between Africanized bees and European bees always result in Africanized offspring. They continue to spread, and ultimately may colonize most of the southern third of North America.

Africanized bees are problematic because, compared to European bees, they are hyper-aggressive. African bees have to be aggressive. They live in a region of the world where the climate can be unpredictable, and where other animals, notably humans, but including creatures like honey badgers (Fig. 19.18), have been robbing their nests for millions of years. African bees defend a much larger radius around their nest, are much easier to rile, and attack in much larger numbers when they perceive the colony is under attack than would European bees. These are the factors that make them dangerous. African and Africanized bees are not larger or more venomous than European bees (their venom

**Figure 19.18.** The honey badger.

There's a Fly in My Soup: Stored Product and Urban Pests

is identical in composition and quantity), but when they attack, they attack much more vigorously. They pose additional problems for beekeepers since they tend to abscond (pick up and leave their hive) much more readily than Europeans, and swarms are likely to settle in inaccessible sites more often. They are also less efficient pollinators and store less honey than Europeans. While they are aggressive around the hive, and they define "around the hive" much more broadly than Europeans, away from the hive they are no more aggressive or dangerous than the others.

Since their original, accidental release in Brazil, Africanized bees have killed one or two people a year in the Americas, on average, and the rate has actually declined as people have learned to live with them. But the idea of perishing from the stings of thousands of tiny insects is a frightening one, and since their introduction, Africanized bees have been the inspiration for a number of highly improbable movies and countless sensationalized media accounts. While these may have some entertainment value, they have served to scare people all out of proportion to the actual threat posed by these insects. Africanized bees certainly are a serious issue, but as millions of Africans can attest, people can learn to live with them.

# The Only Good Bug Is a Dead Bug!
## Insect Pest Management

While the vast majority of insects are beneficial components of the ecosystems that sustain us, as we've seen in the last couple of chapters, a number of them pose significant threats to our wellbeing. Managing pest insects is a sometimes daunting balancing act, as we try to reduce the impacts of these animals while causing as little damage as possible to the ecosystem. Insect populations are extremely dynamic. Insects have tremendous reproductive potential and are capable of moving long distances. One of the most difficult tasks in pest management is simply figuring out when it is necessary to control populations.

We introduced the concept of the economic injury level (EIL) in an earlier chapter (Fig. 18.9). This is an extraordinarily useful tool for identifying when direct pest population measures are warranted. The EIL relies on high quality information derived from careful, frequent observation of potential pest populations, and this requires sampling, since it is impossible to count every individual insect in, say, a crop field. A great deal of research effort goes into developing effective and efficient sampling programs for pest populations, and a fairly substantial industry has grown around pest sampling in agriculture, homes, and businesses.

Modern pest control is built around the concept of Integrated Pest Management (IPM), which relies on the EIL concept to help ensure the most effective control with the lowest impact. IPM systems seek to integrate all possible useful control strategies to reduce pest populations to the point where they no longer cause us economic harm (defined broadly). In IPM, we use those strategies that we know to be environmentally benign first, and resort to harsh control strategies only as a last resort.

We broadly divide pest management tactics in to non-insecticidal and chemical insecticide categories. We'll discuss non-insecticidal strategies first.

## Physical Pest Management

Physical pest management involves manipulating the physical environment to make it less habitable for the target pest. Perhaps the most important and widely used physical tactic is manipulating temperature. As we established in Chapter 7, insects have temperature limits on their activity and development. Most insects cease all activity at temperatures below 40° F, and most die, eventually, with exposure to temperatures below 25° F. On the other end of the spectrum, most insects can't tolerate exposures over about 130–140°F, although there is great range in both heat and cold tolerance among different species. Refrigeration helps keep food fresh by limiting the activity, growth and development of insects, fungi, and bacteria that might already be present in and on the food. Heat treatment has become one of the industry standards for treating bedbug infestations. While it can be extremely effective, it is critically important that all potential bedbug harborage receives elevated temperature. Heat treatment is also used for eliminating wood-destroying insects in lumber and some structures.

The atmosphere of contained spaces can also be manipulated. The atmosphere contains about 78% nitrogen and 21% oxygen. If the oxygen content of the atmosphere in a greenhouse, grain bin, or other space can be substantially reduced, any insects living in that space will perish. Carbon dioxide or nitrogen is frequently used to accomplish this, either in special chambers, impermeable bags, or, sometimes shipping containers or similar vessels. In order for this tactic to work, the altered atmosphere must be maintained long enough to kill all the target pests, and since this interval can vary among species, understanding the biology of the target pests is critical. This strategy leaves treated products unchanged.

Some attention has been given to using sound to manage pest populations. While a large number of insects do have structures ("ears") for perceiving some sounds, most insects don't, and so a great many of the devices marketed to homeowners to provide "ultrasonic" pest control probably don't accomplish much. Sound detection devices, while not actually providing any direct control, can be very useful surveillance tools for the detection of some pests in certain circumstances.

## Mechanical Pest Management

Mechanical pest management is perhaps the most ancient tactic of all. In mechanical management, either the hands or a purpose-built device is used to effect control of the problem insects. We have always used our hands (and sometimes our feet!) to capture and eliminate insects and other arthropods annoying us. Our closest relatives spend long hours engaged in the same behavior (Fig. 20.1), and for some pest situations, this is far and away the most efficient strategy. Hand-picking hornworm caterpillars (Fig. 20.2) and eggs from your garden tomatoes is simpler and less draconian, yet just as effective as getting out the insecticide sprayer. (Of course, this same strategy probably isn't feasible for a farmer with several thousand tomato plants on a couple acres of land).

**Figure 20.1.** Chimpanzee mother grooming her baby for parasites.

174   Insects and People

Perhaps the quintessential mechanical pest management device is the flyswatter (Fig. 20.3)—a simple, but extremely effective tool made for one purpose (although my mom found another use for hers). Mechanical devices go far beyond the flyswatter. A huge diversity of traps, including things like flypaper, light-baited flea traps, and baited roach and ant traps can effectively reduce pest insects. There are also many different vacuum devices, ranging in size from home pest versions to tractor-mounted, commercial agriculture-scale machines.

One of the most well-known devices is the "bug-zapper," made famous in the movie *A Bug's Life*. Bug-zappers (Fig. 20.4) are essentially light traps with an electrified grid that causes insects that impinge on them to, well, explode. The target insects are supposed to be the nocturnal biting flies (read, "mosquitoes") that bedevil night-time outdoor activities. However, female mosquitoes interested in seeking blood meals aren't attracted to light. They're attracted to $CO_2$, body heat and the body odors of their potential hosts, so bug zappers actually end up killing mostly harmless, non-target insects like moths and beetles. Since some of the fragments of blasted insects can travel many feet from the trap, bug-zappers can contaminate food and other products located near the trap. Some of the "pepper" on your potato salad may not, in fact, be pepper, if the salad bowl is near a zapper. Within the last decade, a number of companies have manufactured electric mosquito traps that incorporate a $CO_2$ emitter, and these are clearly more effective at killing mosquitoes, but it's important to remember that they only attract mosquitoes that are downwind of the trap.

Perhaps the most widely used mechanical insect management devices are window screens and other barriers. While screens do keep blowing leaves and the occasional sparrow out of buildings, they are used primarily to exclude insects, particularly flies and biting insects. The widespread use of window screens was critical to the successful eradication of malaria as an endemic disease in North America. The air curtain one occasionally encounters at supermarket entries is another effective barrier, since most insects will not attempt to fly through the strong downdraft produced by the machine. Still another example of an effective mechanical barrier is the plastic wrap used to protect meat and other foodstuffs in the grocery.

**Figure 20.2.** Tobacco hornworm caterpillar on a tomato plant.

**Figure 20.3.** A fly swatter using a flyswatter.

**Figure 20.4.** A bug zapper.

# Cultural Pest Management: Little Changes Can Mean Big Pest Reductions

Farming is a complex activity, demanding constant decision-making and fraught with a multitude of risks, with pest management just one major consideration. Cultural pest management is a suite of practices which involve changing things that one would do anyway in a manner that reduces pest impact, and cultural tactics are the base of most IPM programs in agriculture. Cultural strategies can fall into several major categories, and some of these practices are important in arenas separate from agriculture. Understanding pest biology is critical to the successful use of cultural pest management.

**Figure 20.5.** Cotton plants growing as volunteers in a sunflower field.

**Figure 20.6.** Calves in mud and manure.

Crop manipulation strategies are common in most production systems. Farmers can change planting or harvesting timing to avoid pests, or modify row spacings or seeding rates to make crop fields less attractive to insects. A very important manipulation is crop rotation—not planting the same crop in the same field in successive cropping cycles can starve latent pests, or make it difficult for colonizers to find the crop. It is often important to consider what kinds of habitats are adjacent to the crop field, since many insects migrate from other crops, or other kinds of habitats, into crop fields.

Sanitation strategies are also critically important to effective pest management, both on the farm and in the home. Reducing the resources insects need is virtually certain to reduce target insect populations. In agriculture, sanitation may involve prompt destruction of the residues left in the field after harvest, or the destruction of culls left after sorting fruits and vegetables. Destruction of volunteers, plants originating from waste seed from the last crop (Fig. 20.5), prevents pests on that crop from bridging across a rotational pattern. In livestock operations, proper management of animal waste is essential to reducing flies and other nuisance insects (Fig. 20.6). Around the home, sanitation is important in reducing populations of roaches and other household pests. Leaving dirty dishes in the sink overnight provides a smorgasbord for any roaches living in your kitchen.

Some other cultural strategies are important in agriculture. For many pests, destroying alternate host plants for pests that are growing close to crop fields can greatly reduce those pests. Mowing vegetation near the field or under the canopy of an orchard can help alleviate pest pressures. Proper water management in irrigated agriculture can also help reduce pests. Trap crops, small areas of the field planted to more preferred hosts, can be used to concentrate pests so that they can be destroyed with insecticides or other tactics.

In most production systems, cultural strategies are critically important, but in many cases, they alone won't reduce pest populations to acceptable levels.

# Crop Plant Self Defense: Host Plant Resistance

While we typically grow modern crops in uniform stands of a single genetically uniform variety, all our modern crops originated from genetically highly diverse wild ancestors, and, over the last 10,000 or so years, we have developed large numbers of local "land races" of all our major crops. All this diversity allows modern plant breeders to seek genes that allow one strain of crop or another to resist the depredations of some insect pests. Breeding these traits into standard crop varieties thus allows the plant to protect itself. Such **host plant resistance** (HPR) can be an extremely effective, environmentally benign, and economical pest management strategy.

HPR mechanisms fall into three main categories. **Antibiotic resistance** relies on traits that, in the plant, negatively affect the growth, survival, or reproductive capabilities of the pest insect. Plants produce two major kinds of chemicals: primary metabolites, which are necessary for the growth and development of the plant, and secondary metabolites, which aren't essential for growth but typically protect the plant from one kind of an enemy or another. Most antibiotic resistance mechanisms involve secondary chemicals that the plant produces, although some are physical traits, like leaf hairiness or fruit shape. Antibiotic resistance is only useful if it doesn't affect beneficial insects and doesn't reduce the utility of the crop.

**Nonpreference** or **antixenosis** involves traits that make it hard for the pest to find and identify the crop plant. These traits may involve changes in secondary plant chemicals or the structure of the plant. Nonpreference is only useful if it holds up when the insects have no choice but the crop plant.

**Tolerance** traits allow the plant to withstand the same levels of pest pressure as non-resistant varieties without yield loss. Tolerance is often under the control of complex genetics that are hard to breed in to standard varieties, but once it is achieved, it tends to be very stable, since, unlike the other two forms of resistance, it imposes no selection pressure on the pest that might elicit resistance in the pest.

In order to conventionally breed resistant plant varieties, the genes for those resistance traits must be present somewhere in the crops genome, and sometimes, they are not. The most recent development in HPR is harnessing the new tools of **genetic engineering** to insert genes from species that could never be conventionally crossed, producing transgenic crop varieties. To date, all the genes that have been exploited for insect pest management in this way have come from an interesting soil bacterium called *Bacillus thuringiensis*, or **Bt**, which we will discuss at length a bit later. This technology has been used to produce varieties of corn, cotton, and some other crops that are protected against some of their most important pests, but there is a great deal of controversy over this technology in the general public. Some feel that inserting genes that are so foreign could cause great problems, but it is indisputable that these insect control traits have greatly reduced the use of conventional insecticides in these crops, which is probably a net positive for food safety and the environment.

# Biological Control: Harnessing Our Greatest Allies

All living things are confronted by a host of agents that might kill them. These agents fall into two broad categories: abiotic, or non-living factors like the weather, and biotic or living organisms like predators, parasites, and diseases. In pest management, we call the biotic agents **natural enemies,** and all three of the classes we just mentioned can be valuable.

**Predators**, of course, are animals that hunt and consume other animals for sustenance. For most pest insects, the most valuable predators to us are other arthropods, primarily other insects, and spiders. Amongst the insects, significant numbers of economically important predators are found in the Coleoptera, Diptera, Hymenoptera, and Hemiptera, although other orders also contain some predators. Arthropod predators of other insects range from those that are monophagous on a single species of prey, to polyphagous species that eat pretty much anything they can catch and subdue, including other beneficial insects. In most cases, insect predators have longer lifespans and lifecycles than their prey, and they often have lower rates of reproduction. However, many are behaviorally plastic and can exhibit a functional response to changes in the abundance of different prey species that allow them to effectively reduce pest populations.

Parasites are organisms that live in or on others, stealing the nutrients they need for their growth and development from their hosts. Most parasites would just as soon their hosts never know they were there, and, for most, it's really important that their host not die as a consequence of the parasite's presence, since that usually means their death, too, unless they can quickly find another host. The most important parasites of pest insects, however, are special parasites called **parasitoids**. Parasitoids differ from conventional parasites in that they typically kill their host as part of the completion of their lifecycle. The largest numbers of parasitoids of pest insects come from the orders Diptera and Hymenoptera (Fig 20.7). Most Hymenoptera, in fact, are parasitoids on other insects. However, parasitoidal insects are found in several other orders, as well. Only the larvae of most parasitoids are actually parasitic. They feed on or in their host, usually consuming it completely, save for its exoskeleton, and then pupate. The adults typically feed on nectar or other carbohydrate sources, if they feed at all, and mated adult females are the only host seeking stage. Parasitoids typically are adapted to one or a few closely related host species, and therefore are generally more specific than predators. They also tend to have shorter life cycles than their hosts and can effectively respond numerically to changes in pest density.

*Photos courtesy of Clyde Sorenson*

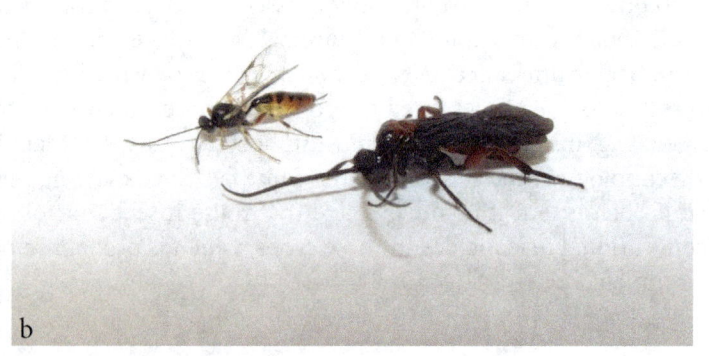

a

b

**Figure 20.7 a.** A tachnid fly. The larvae of tachinids are parasitoids of other insects. **b.** Two wasps that parasitize the tobacco budworm caterpillar as larvae.

178    Insects and People

**Pathogens** are the organisms that cause disease in other species, and the pathogens of interest to pest managers are specific to pest insects. They typically are essentially harmless to all other living things in the ecosystem. Economically important pathogens of pest insects include viruses, bacteria, protozoa, and roundworms (Fig. 20.8), among others. Some species of all these have been commercially developed at one time or another for use as **microbial** or **biological insecticides**, which can be applied to crops or other commodities through a sprayer or other applicator, just like conventional insecticides. Many pathogens require specific environmental conditions to effectively infect their host pest species and cause mortality, and this can be a limitation to their effective use, but naturally occurring **epizootics**, or epidemics, of insect pathogens frequently cause the collapse of pest insect populations with no assistance from humans.

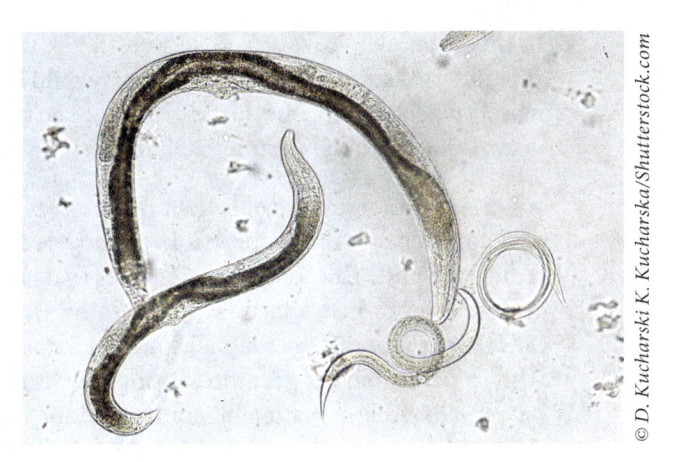

**Figure 20.8.** Nematodes (roundworms) that infect insects.

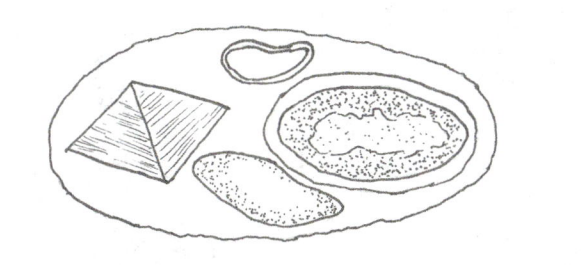

Bt is, for a number of reasons, perhaps the most important insect pathogen in agricultural pest management. Bt is a soil-inhabiting bacterium that is almost ubiquitous, and hundreds of different strains from all over the world have been identified. Bt cells form a tough, stable, long-lived, spore stage when confronted with inhospitable environmental conditions, and, when they enter this spore stage, they form a crystalline,

**Figure 20.9.** A schematic diagram of a *Bacillus thuringiensis* spore. The double-pyramidal structure is the crystalline protenaceous parasporal toxin that causes death in susceptible insects.

protenaceous parasporal body (Fig. 20.9), which is toxic to some insects. Bt kills insects when the parasporal body interacts with the insect's gut lining, causing perforations in the gut and massive septicemia from other gut bacteria invading the body cavity. Each strain of insecticidal Bt (some are not) produces a parasporal body toxic to only one or a few species of insects. Most are toxic to different Lepidoptera, while a few are toxic to certain beetles, and others, to certain flies. A strain that affects one kind of caterpillar typically has reduced or no toxicity to other species of caterpillars and this great diversity of strains allows scientists to develop specific commercial strains for specific groups of pests. Different strains of Bt have been used as effective microbial insecticides for caterpillars, beetle larvae, and in mosquito control for several decades. Once a suitable strain has been identified, it can be grown economically in bacterial culture, and the spore stage gives the product a long and stable shelf life. The bacteria don't have to be alive to be effective since it is the parasporal body that actually kills the target insect. However, Bt only kills target insects when they eat it, and while the spores are tough, sunlight can degrade the parasporal body relatively quickly if the spore is exposed to the sun on a leaf. This short residual life is one of the greatest limitations to the use of Bt as a microbial insecticide and was one of the main motivations to using the genes from Bt to create the transgenic crop plants we previously mentioned. In that application, only those insects that actually feed on the transformed plant tissue are exposed to the Bt, and, since the insecticidal protein is protected inside the plant tissue, it isn't degraded by the environment.

Biological control agents can be exploited in three main ways. By far the most common strategy is **conservation biocontrol**, which is pretty much what it sounds like—we try to preserve and encourage the full suite of naturally occurring natural enemies that already inhabit the agroecosystem.

In this approach, if we must use a pesticide, we select one that has the least impact on the natural enemies present. We might also preserve wild habitats near the crop field that harbor the beneficial insects when the crop, or the target pest, aren't present. Conservation biocontrol might also include providing additional nesting habitat for some natural enemies.

**Augmentation biocontrol** is the practice of releasing additional individuals of a particular species of natural enemy that is probably already present, but at population levels too low to effectively reduce the pest. This, of course, requires that large numbers of the enemy to be released be raised in an **insectary** or insect farm. It is generally far cheaper and easier to raise large numbers of parasitoids than it is to raise predators, since predators, as they grow, require a large number of prey items, while a parasitoid only requires a single individual of the host, so most augmentation is done with parasitoids. Augmentation biocontrol is done with the understanding that once the pest population is suppressed, the released enemy population will probably die off, and the process will have to be repeated should the pests build back up again. In this respect, augmentation is somewhat similar to using an insecticide. Successful augmentation requires the production of very high-quality natural enemies that can be distributed across the crop field in a safe, efficient, and economical way. These challenges have limited, to some extent, the deployment of this strategy.

The third strategy is called **introduction biocontrol** or **classical biocontrol**. In this case, we are usually trying to manage an exotic pest which is not well suppressed by native natural enemies. Exotic pests usually invade new lands unaccompanied by all of the natural enemies that helped regulate their populations is their homeland. Scientific research on these natural enemies in the exotic pest's land of origin allows us to evaluate their potential impact on the pest. We then import some of these natural enemies, evaluate them to make sure they won't harm harmless, native relatives of the pest (or cause any other kind of mischief), rear them in large quantities in insectaries, and then release then into pest populations. If this is successful, the result is a permanent reduction in the pest population to levels well below the EIL.

**Figure 20.10.** Cottony cushion scale.

**Figure 20.11.** The Vedalia beetle, a biocontrol agent for the cottony cushion scale.

One of the first examples of successful introduction biocontrol involved a pest called the cottony cushion scale (Fig. 20.10), which threatened to wipe out the nascent citrus industry in California in the late 1800s. One of the first economic entomologists employed by the United States Department of Agriculture, Charles Valentine Riley, recognized that the scale wasn't native to North America and that it was probably native to Australia, the source of many ornamental plants newly introduced to California. He sent a gentleman named Albert Koebele to Australia to study the natural enemies of the scale in its native range. Mr. Koebele identified several parasitic wasps and other natural enemies, but determined that a small ladybird beetle called the Vedalia beetle (Fig. 20.11) was the most promising species. Vedalia beetles only eat cottony cushion scales. Over six months, he sent 514 vedalia beetles to California, where they were liberated in the infested orange groves, and within two years, as the beetles increased and were redistributed to additional growers, the scale ceased to be a serious threat to the citrus industry. Vedalia beetles continue to provide excellent suppression of cottony cushion scales across America.

While there have been many other, similar successes in classical biocontrol for other exotic insect pests, many introductions have had little or no effect. Classical biocontrol has also been an important strategy for reducing introduced weeds. Over the last hundred or so years, however, we have learned that it is critically important to make certain that introduced biocontrol agents don't harm native organisms. Some native insects, like the spectacular giant silkworm moths, and some native plants, including some rare cacti, have suffered population declines due to inadequately screened, introduced, natural enemies. Some introduced enemies have become problems in other respects. The Asian multicolored ladybird (Fig. 20.12), introduced to control aphids in nut orchards, has displaced native ladybird beetles and is also a significant nuisance for homeowners whose houses are invaded by thousands of the insects in the fall, when they seek places to spend the winter.

**Figure 20.12.** Asian multicolored lady beetle.

# The Last Gun We Draw: Insecticides for Pest Management

Pesticides are substances that are used to reduce the impact of one kind or another of organisms that cause us problems. Pesticides are diverse; herbicides are used to control unwanted plants, fungicides, unwanted fungi, piscicides, unwanted fish, and so on. Insecticides are just the pesticides used against insects and other arthropods. Pesticides have an ancient history but have been very highly developed over the last 60 years or so as our understanding of chemistry has flourished. Worldwide, about five billion pounds of pesticides, worth about $50 billion US are made and used. The United States is responsible for about one-fourth of the global total. Pesticide use is highest in the developed parts of the world, like North America, western Europe, and eastern Asia, where labor is expensive, populations are dense, and the societies are relatively affluent. In developing parts of the world, where labor tends to be much cheaper, the tasks performed with pesticides in the developed world are done largely by hand.

The biggest share of pesticides made and used, both globally and in the United States, is composed of the herbicides used to control weeds (Fig. 20.13). In the United States, insecticides are about one-fifth of the total and represent the second largest class of pesticides. Insecticides are probably best classified by their **mode of action**—how they interact with the **target sites** inside the insect to kill it.

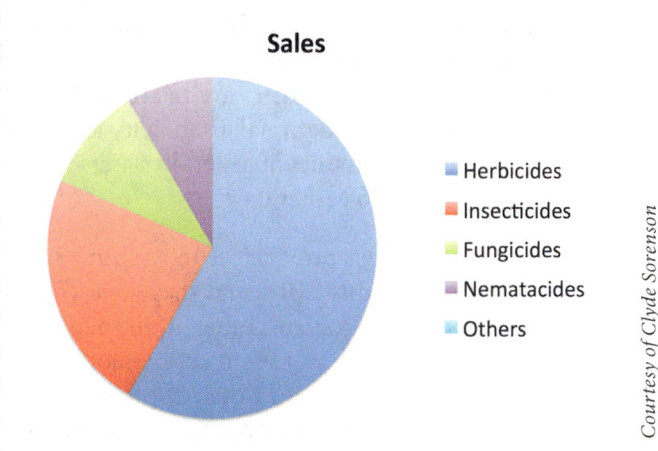

**Figure 20.13.** Pesticide usage percentage in the United States by class, 2007.

Perhaps the oldest target site is the insect cuticle, and the oldest mode of action, cuticle toxicants. The insect cuticle is waterproofed by its outermost layer, the epicuticle, and its primary function in most insects isn't to keep water out, but to keep water in. Anything that compromises the epicuticle's ability to do this is likely to result in the death of the insect due to dehydration. Some ancient insecticides, like wood ash and some dusts, do this by scratching the epicuticle. The epicuticle can also be damaged by soaps and detergents that cause the epicuticular waxes to clump. Still other, more recent, cuticle toxicants work by interfering with the insect's ability to manufacture chitin as it builds a new exoskeleton in preparation for molting. Such pesticides tend to be very safe for mammals and other vertebrates since we don't have insect-like cuticles.

Another rather unique target site is the insect respiratory system. Occluding the spiracles of the tracheal system prevents oxygen from reaching the internal organs and eventually kills the insect. Water is an ancient insecticide that has been used to either kill or exclude at least some insects in crops like rice. Oils can also be useful respiratory toxicants. Heavier, thicker oils are used in orchard crops during the trees' dormant season to kill insect eggs and overwintering scale insects. Lighter, thinner oils can be used on some crops during the growing season for these and other small insects. Respiratory toxicants that occlude the spiracles tend to be quite safe for non-target vertebrates. Fumigant chemicals, on the other hand, can be quite hazardous, and thus can only be used by specially trained applicators. While the route of entry for a fumigant is respiratory, the chemicals have a number of effects on different tissues in the insect and are generally regarded as narcotic in their action on the insect. They also typically kill weed seeds and plant pathogens as well. Once a treated area is ventilated, there is no residue of the fumigant, making these materials useful for treating stored products, harvested produce, and soil.

The insect endocrine system is one of the newer target sites, and is also one of the safest for non-target vertebrate organisms, because of the vast differences in arthropod and vertebrate hormones. Insecticides that mimic juvenile hormone keep susceptible immature insects immature, and can be useful for insects that are only pestiferous as adults, like fleas and mosquitoes. Some of these materials can also have effects on the reproductive abilities of some adult insects. Insecticides that mimic molting hormone cause immature insects to molt prematurely and incompletely since the insect doesn't have the time to develop a new, functional cuticle. The molting hormone mimics that are currently available work primarily against caterpillar pests.

The insect nervous system is by far the most common target site for most modern insecticides, and most work by either over-stimulating the nervous system until it is exhausted, or by interfering with nervous system receptors so that signals can't be transmitted. Insecticides that shut down the nervous system have one huge advantage over most other insecticides and that is the rapidity with which they take effect. If the nervous system quits working, all other life systems do, too. However, while there are some significant differences between insect and vertebrate nervous systems, there are also many important similarities, and so some nervous system toxicants are quite hazardous to vertebrates, while others are relatively low in toxicity. The trend over time has been towards a decline in average mammalian toxicity along with tremendous increases in insect toxicity, so that much smaller amounts of generally safer insecticides are now required to achieve acceptable control, compared to earlier days.

While insecticides provide the most immediate response of virtually any of the pest management strategies we've discussed, and they typically provide very high reductions in pest numbers, there are a number of problems inherent in their use that relegate them to the role of tools of last resort. Virtually all, including the most benign, produce some collateral damage by killing beneficials, and many are so broad in their spectrum of activity that they kill most all insects present. They almost invariably produce only a temporary reduction in pest numbers, often resulting in a need for additional applications. Because of their very high efficacy, insecticides impose very strong selection pressure on pest populations to develop resistance to these agents, by removing the susceptible individuals and leaving only those rare individuals that can withstand the insecticide. Once resistance is fixed in a population, that insecticide and all that work in the same way are no longer useful. Some can also be quite hazardous for applicators and farm workers to work around. In spite of all these potential problems, insecticides are extremely valuable tools in pest management, and they usually can be used safely and effectively, with appropriate foresight and care.

Managing pest insects will always be a critical task in agriculture, around our homes, and in our businesses. It will be interesting to see what new, effective, and environmentally acceptable strategies we'll develop in the future.

# XXI

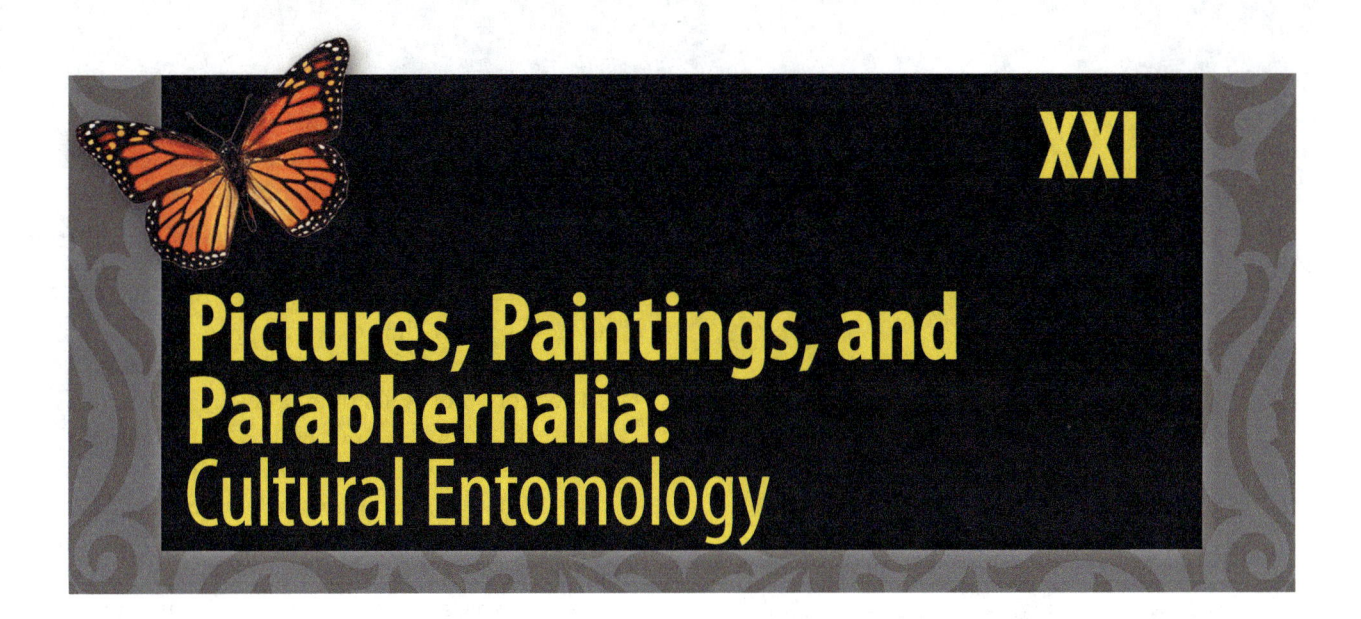

# Pictures, Paintings, and Paraphernalia: Cultural Entomology

We've learned that insects are all around us in our environment. They are clearly among the most abundant animals in most natural ecosystems, and they are even quite common in our homes. But they are also incredibly abundant in our **cultural environment** as well. Insects abound in our literature, arts, and entertainments, and in our religions, histories, and laws.

The study of the occurrence and influence of insects in and on our cultural environment is called cultural entomology, and it is a very diverse discipline, indeed. The *Digest of Cultural Entomology* was published during the 1990s, under the editorship of Dr. Charles Hogue, who was a curator at the Los Angeles Museum of Natural History. Sadly the journal died with his passing. However, during its brief run, the journal published articles on over 60 different aspects of the interactions with and influences between insects and our cultural environment. In this chapter, we'll only be able to discuss a few of these areas at any length—just remember that there is much more!

In almost all the diverse arenas of cultural entomology, it is important to recognize the incredibly intense emotional responses many insects almost universally elicit. Most folks find cockroaches repulsive and butterflies beautiful, and there are a great many other, similar associations. The cultural result of these associations is that many insects have come to have symbolic meaning that can be exploited in various cultural environments. We'll discuss several specific examples as we proceed.

## Six-legged Gods: Insects and Religion

Since time immemorial, people with time on their hands have observed the behavior of the insects around them and tried to come up with explanations for what they see, and since we achieved consciousness, we've tried to explain and understand our place in the cosmos. Frequently, these efforts have intersected, and insects have become integrated into our resulting religious understandings. One of the most well-known examples of these kinds of relationships occurred in ancient Egypt with one beetle that has, to us at least, a rather unsavory lifecycle: the dung beetle (Fig. 21.1). The dung beetle in question is one of the ball rollers. It scoops together a mass of the material, forms it into a compact ball, and then rolls it some distance from the original deposit before burying it in the ground to protect it from other dung-feeding insects that would steal it. It then lays its egg on the dung ball, and the dung serves as the food resource for the developing larva. Some weeks later, a new beetle emerges from the soil, and the cycle starts anew.

**Figure 21.1.** A roller dung beetle, very much like the scarab venerated by ancient Egyptians.

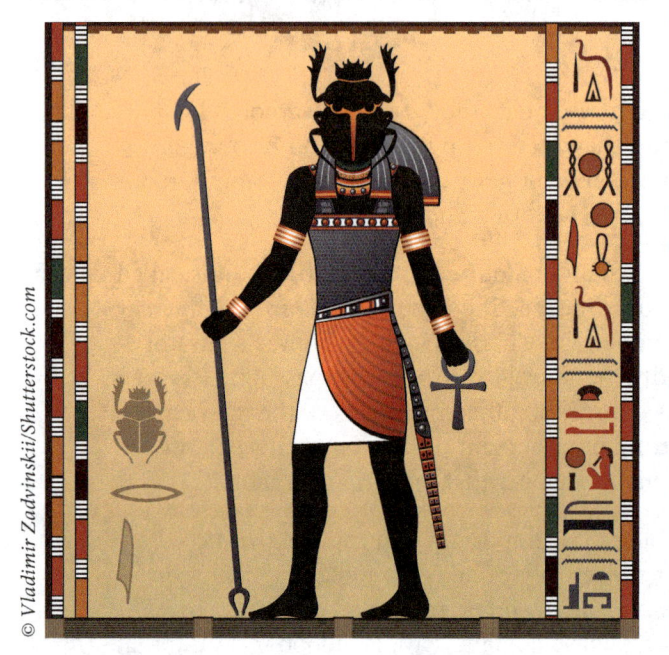

**Figure 21.2.** A representation of Khepera, the chief deity in the ancient Egyptian pantheon.

**Figure 21.3.** A carved scarab.

The ancient Egyptians saw metaphors for their belief in an afterlife in the behavior of the beetle. In their religious construction, their chief deity Ra, assumed the form of Khepera when he rolled the sun across the sky daily, renewing life and recycling souls, much as the dung beetle pushed the ball of dung across the surface of the earth. The new beetles that emerged from the places old dung balls had been buried were metaphors for the rebirth of souls in an afterlife, which tied the insects to their creator god, Atum. Most representations of Ra (Khepera) in hieroglyphics depict him as a scarab-headed man (Fig. 21.2) and carved scarabs were ubiquitous talismans in ancient Egypt (Fig. 21.3). Curiously, other cultures in disparate parts of the world, including South America and sub-Saharan Africa, have observed dung beetles and integrated them into the mythology supporting their religions.

The observation that many other insects go through dramatic transformations has led to them being integrated into religious contexts. Butterflies have represented the souls of the dead and the rebirth of the dead in several Amerindian cultures in North and South America, and represent Jesus in many Christian traditions. They were dream-bringers to some Plains Indian tribes. Beetles and their behaviors have resulted in their integration into several creation myths in different parts of the world. Because insects are important and sought-after food in many places, some species, particularly certain grasshoppers, have acquired significant ceremonial significance.

Beyond actual deification, many insects have important symbolic meanings in different parts of the world. Social insects are almost universally admired (at least when they're not invading one's kitchen) for their apparent industry and their ability to achieve much more together than individually, and so they have become symbols of hard work. Flies, because of their association with filth and disease, are almost universally associated with evil. In China and Japan, crickets are regarded as bearers of good fortune. Many folks in China keep crickets in the house as pets, both for the luck they may bring and to enjoy the songs of the males. In China, cricket fighting is also a significant wagering sport. Crickets have positive associations in other parts of the world, as well.

186    Insects and People

# Insects and the Written Word

It should come as no surprise that insects have shown up in various literary works, pretty much since the beginning of writing. Several of Aesop's Fables involve insects. Perhaps the most famous of these is "The Ant and the Grasshopper," which teaches the lesson that preparing for the future is a wise thing. This story was first recorded about 2,500 years ago, and yet it was the basis for the movie, *A Bug's Life*. In the intervening millennia, insects have been steady fixtures in novels, short stories, plays, and other works of fiction. Insects can assume a number of roles in these works. They can be simple, straight-forward representations of themselves, or function as devices to describe an environment or to enrich a scene. However, they often can take on symbolic meaning of one sort or another through allegorical or metaphorical uses. Insects that we generally have positive attitudes towards are generally going to represent positive things, while those that elicit negative feelings are going to represent dark forces. For instance, in *The Lord of the Flies*, the flies buzzing around the "lord's" head may represent the corrupting effects of anarchy.

Many fantasy and science fiction works feature insects and other arthropods as primary characters or important secondary characters. Both *Bug Wars* by Robert Asprin and *A Taste for Honey* by Gilbert Heard feature giant insects as primary antagonists. *The Swarm* by Arthur Herzog is about a town attacked by a huge swarm of "killer" bees. Giant spiders are particularly common in fantasy and science fiction. Shelob in Tolkien's *The Two Towers* and Aragog in *Harry Potter and the Chamber of Secrets* are two great examples.

A number of very famous works prominently feature insects or other arthropods. Perhaps the most famous of these is Kafka's *The Metamorphosis*, in which the main character, a morose man named Gregor, awakes one day to find that he has been transformed into a giant "verminous monster." The bulk of the tale describes Gregor's attempts to come to terms with his transformation and its effect on his family. Ultimately, he dies to unburden them. In H.G. Wells' *The Empire of the Ants*, a race of super-intelligent ants overrun a Brazilian town and threaten to take over the world.

The best-selling book in the world, the Bible, makes frequent reference to insects and other invertebrates. By one count there are about 120 mentions of these creatures in the Old and New Testaments. Many of these references are mundane—which insects are kosher, or allowed, as food, and which aren't, for instance. Others, however, exploit the ancient stereotypes we have of some that persist to this day. Flies are associated with evil because of their association with filth; locusts are a marauding pestilence. Ants, on the other hand, are role models of industry, and hornets are armed allies of the Chosen People.

Children's literature makes abundant use of insect characters. Insects and other arthropods are inherently interesting to children, and the fact that most undergo change and metamorphosis, somewhat like children, also makes them useful plot devices. Most insect characters in children's books should best be called "bugfolk"—anthropomorphized characters based on insects, often exploiting common stereotypes of different insects. In general, these bugfolk are sympathetic and often humorous characters that assist humans. In many works, they are the main characters. One of the most popular children's books of the last 100 years is *Charlotte's Web*, by E. B. White, which tells the story of a pig saved by a "writing" spider. Several of Eric Carle's books are about insects, with *The Very Hungry Caterpillar* probably the most recognized. In it Carle does an excellent and entertaining job of describing metamorphosis to young children.

Insects have long been important devices in poetry. Poetry demands economy of word, since the goal is to elicit emotion within the constraints of a poetic form. Insects evoke strong reactions, both positive and negative, and their use can intensify the emotional response in the reader. Poems specifically about insects and their short, intense lives can contain strong allegorical messages about the human condition. Of course, some poems about insects are just that—poems about the delight one can experience simply observing insects.

William Shakespeare frequently used insect allusions and references in his sonnets, plays and other works. As in other venues with other writers, Shakespeare exploited the stereotypes associated with particular insects. He, as others, exploited the association between flies and filth, and the supposed industry of ants and bees. Like his contemporaries, Shakespeare erroneously thought social insect colonies were "led" by male "kings," but I suppose he can be forgiven this mistake.

John Milton, a statesmen and poet who lived about 50 years after Shakespeare, and who is the author of *Paradise Lost*, made at least 25 references to insects from at least six orders in his many poems—at least a dozen in *Paradise Lost* alone. Thomas Hood, Who lived in the early 1800s, was perhaps the most effective entomological poet, and seemed more than most, to understand insects and their behavior. A couple of his works contain more than a dozen entomological references. Emily Dickinson wrote at least 46 insect-related poems. Longfellow, Tennyson, Keats, Wordsworth, Ogden Nash, and dozens of others have used insect allegory and imagery in their poems.

Insects also invade our everyday language, again because they provide convenient shorthand for common thoughts and feelings. Everyone understands "He has ants in his pants" as a description for someone (often a child) who simply can't sit still, because it is not hard to imagine how difficult it would be to remain motionless with trousers full of swarming Formicids (of course, some of us, present company included, have actually had this experience and so relate readily to it). Table 21.1 lists a number of these common turns of phrase.

# The Beat(le) Goes On: Insects in Music and Other Performance Arts

Insect themes have been common in music ever since there has been music. They even make appearances in more classical pieces (i.e., *Flight of the Bumblebee* by Rimsky-Korsakov). However, they are perhaps most common in popular music, particularly that of the last 50 years or so. Joe Coelho, an

**Table 21.1** *Some common idioms exploiting insect imagery*

| | |
|---|---|
| Ants in your pants | Buzz off |
| Like ants to honey | Knee high to a grasshopper |
| A bee in your bonnet | I caught the flu bug |
| As busy as a bee | She's a social butterfly |
| The bee's knees | Float like a butterfly, sting like a bee |
| Mad as a hornet | A fly in your ear |
| Don't bug me | Like a duck on a junebug |
| As snug as a bug in a rug | You tick me off |
| Bug off!! | Don't let the bedbugs bite |
| This room is bugged | Nit-picker |
| Butterflies in my stomach | Flea-bitten dog |
| Like a moth to a flame | Quick as a cricket |
| Flea brained | No flies land on a boiling pot |
| Flea-bag hotel | Even a flea can bite |
| Nit wit | A fly in the ointment |
| In a gnat's eye | You can catch more flies with honey than you can with vinegar |

entomologist who studies the physiology of Hymenoptera, did a comprehensive study of insects in pop music in 2000. He identified about 1,400 artists, albums or song tracks that had something to do with insects. Three orders seem to dominate pop music in this analysis: the Hymenoptera, Lepidoptera, and Diptera. Most uses of the first two are positive, because of the positive associations we have with bees and butterflies. Most of the Dipteran associations played off the negative images flies have. Even though beetles are the most diverse creatures on earth, relatively few associations feature the Coleoptera.

One of the most famous bands of all time, however, was named The Beatles—partly in homage to one of their favorite bands, Buddy Holly and the Crickets, and partly because it played on the word "beat" in homage to the Beat movement of the late 50's. A number of other artists have adopted insect-themed stage names. Iron Butterfly were subsequently inspired by the Beatles to incorporate an insect theme in their group's name. Gordon Sumner, aka "Sting," acquired his stage name as a struggling musician in London, because he habitually wore a yellow and black striped sweater. Alien Ant Farm chose their name in recognition of how alien ants are to most folks.

Insects or insect allusions have been prominent in other kinds of performing arts, too. *Madama Butterfly* is an Italian opera about a Japanese girl nicknamed "Butterfly," who is married to an unfaithful American. *Le Festin de L'Araignee* is a French ballet from the turn of the last century by Albert Roussel. The title translates to "The Feast of the Spider." Plays have featured insects from the time of the ancient Greeks. The Father of Comedy, Aristophanes, penned the satirical *The Wasps* in 422 BCE, and insects have pretty much been on the stage ever since. All but two of Shakespeare's plays contain insect references, as do a great many lesser known stage productions. Jean-Paul Sartre, the noted French philosopher, wrote *The Flies*, a commentary on the Nazi treatment of the French during World War II. *The Metamorphosis* has also been produced as a stage play. As we'll see later, insects have been extremely prominent in the most popular performance art of the last 100 years or so—the movies.

## Pictures and Paintings: Insects in the Graphic Arts

Insects have long been popular motifs in decorative arts. Remember, some early artists decorated their ceremonial caves with images of the creatures, in amongst the cave bears and aurochs. The most popular insects in decorative arts tend to be the insects that we find aesthetically pleasing—creatures like butterflies, bees, dragonflies, and some beetles.

Insects, in some ways, are tailor-made to be motifs in jewelry, since they are, for the most part, small, and many have vivid colors (Fig. 21.4). However, real insects have also been incorporated into jewelry. Some modern jewelry makers incorporate resin-protected butterfly wings into their works (Fig. 21.5), while the wing covers of some of the tropical buprestid beetles have also been used to make pieces (Fig. 21.6). Some enterprising artists have exploited the case-making behavior of caddisfly larvae to make their pieces. They provision aquariums containing the larvae with semi-precious stones and allow the insects to incorporate them into their silken cases. The resulting cases can then be filled with resin and turned into earrings or pendants (Fig. 21.7). In Mexico, there is a long tradition (dating back to the Maya) of decorating living Mequech beetles with semi-precious stones and wearing the tethered creatures as animated broaches.

Postage stamps have been around for about 170 years. Prior to their development, the wood stamps used to mark postage could easily be forged, and the onus of paying for a mailed item fell on the recipient, so the development of state-sponsored and regulated postage was a boon to communication. Insects have been on stamps for about 125 years, with the first, not surprisingly, depicting a

**Figure 21.4.** A selection of insect-inspired earrings.

**Figure 21.5.** A pendant made with a morpho butterfly wing, courtesy of Lepidopteran Art.

**Figure 21.6.** A pendant made of jewel beetle elytra.

**Figure 21.7.** A pair of caddis-tube earrings.

**Figure 21.8.** Australian stamps depicting butterflies.

**Figure 21.9.** A Cuban stamp portraying a mosquito.

honey bee. World-wide, approximately 5,000 stamps have been produced with images of insects, although only a relative handful (less than 30 or so) have come from the United States. Far and away the most common insects on stamps are butterflies (Fig. 21.8), followed by bees and other "positive" insects. However, a substantial proportion of stamps has depicted pest insects, like malarial mosquitoes, as parts of national education campaigns (Fig. 21.9).

Insects on coins also have a somewhat interesting history. Coins, as a means of transferring wealth, have been around for about 2,800 years and were probably invented in ancient Turkey. The classical Greeks quickly adopted this technology. Classical Greek society consisted of a number of more or less autonomous city-states, and each made its own coinage. A large number of these classic period coins featured insects: grasshoppers, bees, beetles, moths, and so on. In most cases, the insects were supporting characters in depictions of Greek gods, but in some, the insects were the main motif. Many of these were extraordinarily beautiful (Fig. 21.10). As the ancient Greek society declined with the ascendancy of the Roman society one peninsula to the west, insects started to disappear from coins. By the time Julius Caesar assumed power in 49 BCE, insects on coins largely disappeared. Up until very recently, very few coins minted anywhere in the world depicted insects, but in the last few decades about 100 coins have been produced with these animals on them; however, no US coin has ever depicted an insect, although a beehive almost made it on to the Utah state quarter dollar in 2007.

**Figure 21.10.** A drawing of an ancient Greek coin depicting a honeybee.

Insects have also long figured in paintings, and they often have symbolic significance in these works. During the Middle Ages and the Renaissance, for instance, bees often represented the Virgin Mary, and a beehive, the church. Butterflies often symbolized purity or the fragility of youth and of life, while flies were generally associated with evil. Some artists exploited these associations to make sometimes humorous commentaries on the people they were painting. A fly inconspicuously placed within the framing of a painting might suggest that the artist had a rather low opinion of the sitter, while the inclusion of a butterfly might suggest that the sitter died before the painting was completed. Supernatural figures, such as cherubs or fairies, were often depicted with the wings of Lepidoptera to exploit the symbolic meaning associated with them.

Insects have continued to intrigue painters. Many more recent artists have exploited insects both for symbolic meaning and because the shapes and colors of insects provide visually arresting figures to depict. M. C. Escher included insects in a great many of his works, including *Mobius Strip II*, and *Dream Theater*. Perhaps the most noted surrealist artist of the last 100 years, Salvador Dali also frequently included insects in his paintings, including his most famous, *The Persistence of Memory*. Dali's insects always have symbolic significance, but the symbolic universe he used is largely one of his own creation. He was both fascinated and repulsed by insects, and, for a time, apparently suffered from delusional parasitosis.

E. A. Seguy was a French artist and designer who produced two albums of insect images to encourage other designers to utilize the interesting patterns and vibrant colors the animals possessed. He imparted no symbolic meaning to the creatures. He just enjoyed, and wanted us to enjoy them, as aesthetically intriguing forms. William B. Rowe similarly was interested in insects primarily because of their forms. Rodney Matthews, a British illustrator who has done the artwork for many rock album cover jackets and fantasy books, is perhaps one of the most accomplished bugfolk artists. A number of his works have depicted beautifully rendered but quite fantastic insects, often engaged in making music.

**Figure 21.11.**

Once reserved for sailors enjoying interesting experiences in foreign ports, tattoos are an ancient art form that has increased tremendously in acceptability and popularity in North America and Europe over the last two or three decades. Insects have long been popular motifs for tattoos (Fig. 21.11). There appear to be certain gender biases in the types of insects that seem to appear in tattoos. While Hymenoptera and Coleoptera might show up in tattoos on folks of either sex, butterflies are pretty rare on guys.

Insects have also long been important to cartoonists. A cartoonist, much like a poet, has to elicit a reaction in his or her reader in a very limited space, and insects are often economical devices for generating this response. Many insects produce gut reactions, and others have associated stereotypes that can be exploited. Further, there is greater "poetic license" with insects than there might be with other animals. If something dastardly happens to an insect, it might be viewed as humorous, while the same event happening to, say, a bunny rabbit might get quite a different response.

Perhaps the greatest "insect cartoonist," at least in the estimation of most entomologists, is Gary Larson of *The Far Side* fame. Even though he has been retired since 1995, his cartoons continue to decorate the office doors and bulletin boards of entomologists the world over. Mr. Larson attended Washington State University majoring in communications, but he grew up with a deep love for natural history, and his understanding of biology and animal behavior is probably the main thing that makes his cartoons so enjoyable to so many scientists. Entomologists have recognized his contributions to the collective sanity of the entomological community by naming at least two species of insect after him: *Serratoterga larsoni*, which is a tropical butterfly, and *Strigiphilus garylarsoni*, a chewing louse found only on owls. Unfortunately, the butterfly has been renamed due to the nomenclature rules of systematics, but the owl louse remains. The entomologist who described and named *S. garylarsoni* was uncertain as to whether or not Mr. Larson would consider having a parasite on a "creature of the night" named after him a positive or negative thing, so he wrote a letter requesting his permission. As it turned out, Mr. Larson was thrilled to have the creature named after him, and so it stands.

We have just dipped a toe into the topic of cultural entomology. As we've demonstrated, though, insects have had significant and diverse impacts on our cultural environment, and they most certainly will continue to in the future.

# XXII

# The Hellstrom Chronicles: Insects in the Movies

In the previous chapter, we briefly discussed the role (pardon the pun) of insects in performing arts. Insects, or, at least, the idea of insects, have been incorporated in performances of one kind or another almost since the beginning of human civilization. With the advent of celluloid movie film, projection technologies, and special effects, the opportunity to create ever more elaborate and fantastic tales emerged, and the integration of insects and other invertebrates into visual story telling blossomed. Insects have been in the movies since their beginning, and a number of distinct genres of insect-themed movies have emerged over the last 120 years or so.

## Before Surround Sound: Insects in Silent Movies

The first movies were silent movies, projected in a theater, usually with a live music accompaniment, with the story driven by the visual images and periodic title cards. One of the more remarkable early efforts was a stop-action animated movie by the Russian Ladislaw Starevitch called *The Cameraman's Revenge*. Starevitch was one of the earliest filmmakers to use the stop-action animation technique, in which models are moved slightly for each frame to generate the movement of the characters. The movie tells the tale of a philandering married couple, and the characters they encounter in their affairs are all "played" by insects. The insect characters are actual specimens manipulated through stop-action. He also made *The Insects' Christmas*, which tells the story (in six and a half minutes) of a Father Christmas tree ornament that comes to life and constructs a Christmas tree for insects and other "forest people."

Many other silent movies that incorporated insects used either real insects or mock-ups of real insects as props that the human characters responded to, often for humorous effect. A good example of this type of movie was *Bumbles Goes Butterflying*, where the main character single-mindedly pursues a butterfly through all sorts of places. Bees and hornets were frequently used for comic effect in silent movies, as well.

# Jiminy Cricket and Friends: Insects in Animated Movies

As we've seen with Starevitch's works, insects entered animation early on. With the emergence of sound recording technologies and cel animation (in which drawn and painted characters on transparent cels are placed sequentially over stationary painted backgrounds), the number of insect characters exploded on screen. Most animated movies target children, since, as we've mentioned previously, insects are inherently interesting to children. Insect characters in animated movies are typically highly anthropomorphic—that is, made to take on human characteristics, and, thus, are basically animated bugfolk. Perhaps the most famous of these characters is Jiminy Cricket, from Walt Disney's 1940 *Pinocchio*. A year later, another orthopteran, this time a grasshopper, was the star of *Hoppity Goes to Town*, which tells of the return of the main character to a bug-inhabited city nestled in amongst the "real" city.

Insect characters populate many animated feature movies, but usually as supporting characters, as in *Pinocchio* and with Cri-kee in 1998's *Mulan*. However, there have been several more modern movies that "star" insects and other arthropods. In general in an animated movie, the more anthropomorphic an arthropod character is, the more sympathetic that character generally is. *Charlotte's Web* was produced as a traditional animated movie in 1973, and is a pretty faithful translation of the book. Of course, one of the main characters is the spider, Charlotte. Charlotte is clearly a spider, but she is depicted with a kind face and voice, and is therefore a very positive character. In 1995, Roald Dahl's book, *James and the Giant Peach,* was made into a very successful computer animated/ live action feature. *James and the Giant Peach* is notable in that the animated arthropod characters are fairly realistic, yet they remain sympathetic characters.

The year 1998 saw two blockbuster animated insect movies: Pixar and Disney's *A Bug's Life* (based loosely on Aesop's fable of the Ant and the Grasshopper), and Dreamwork's *Antz*, which tells a tale of eugenics in an ant colony. These two films present an interesting contrast; while both made money, *A Bug's Life* was far and away the more commercially successful, and continues to be extremely popular. *A Bug's Life* was clearly targeted to the youth market, while *Antz* was, oddly for an animated movie, actually targeted at an adult audience, and the depictions of the ants reflect this. While both movies got some important aspects of ant biology wrong (both, for instance, depict male workers as the protagonists), the differences in their ants are dramatic. The ants in *A Bug's Life* are far more anthropomorphic than those in *Antz*. They are blue, four-limbed rather than six-limbed, and have highly stylized faces on huge heads with great big eyes, resembling children in some respects. Those in *Antz* are a nice, anty brown, have three distinct body regions, six limbs attached to the right body part, and smaller, more anty heads and faces; *Antz* also does a better job of depicting ant metamorphosis, with juveniles that are actually larvae, rather than just miniature versions of the adults. The end result is that the ants in *A Bug's Life* are far more sympathetic than those in *Antz*. It's also worth noting the relative anthropomorphization of the villainous grasshoppers in *A Bug's Life*, compared to the ants. While the ants are highly anthropomorphic, with four limbs and smooth "skins," the grasshoppers are much more buggy, with six legs, three body regions, and platy, armor-like exoskeletons, making them far less sympathetic. The "alienness" of the grasshoppers makes it easier to accept the bad things that happen to the grasshoppers at the end of the movie (sorry for the spoiler…).

*Bee Movie* (2007) was a bit of a mess of a movie, with a plot based on interspecific conflict and attraction between bees and humans. Its insects were also highly anthropomorphized, with only four limbs and large, human-like heads, and the makers of this movie, too, got bits of insect biology wrong (i.e., male worker bees, and male bees with stingers). Still it was moderately entertaining and fairly successful. Another movie that came out about the same time, *The Ant Bully*, also played off of insect-human relationships. It's a sure bet that insects will continue be featured in animated movies.

# Improbably Large: The Big Bug Movies

The end of World War II was delivered by an extremely deadly new technology—the atomic bomb. While most were certainly glad to see the end of the war, this Pandora released from the nuclear box created a tremendous amount of anxiety amongst the public. Just what had we done by exploiting this terrible new weapon? What might the consequences of continued development of nuclear technology and the increased radiation exposure it might create, do to our world? An outgrowth of this apprehension was the emergence of the "Big Bug" movie.

During the 1950s and early 60s there was a string of science-fiction movies in which a community was threatened by giant arthropods of one sort or another. Usually, these fearsome beasts were the result of exposure to either radiation from nuclear research gone awry, or some chemical cocktail (another anxiety-inducing emerging technology of the mid-1900s). Several of these are considered sci-fi classics (although some are classic because they are so bad) including *Them*, (1954) featuring giant ants; *Tarantula!* (1955) with a 100 foot spider; *Deadly Mantis* (1957) in which a giant preying mantis is liberated from a glacier (?!); and *Mothra* (1962) featuring, not surprisingly, a giant moth. Giant bugs have remained a staple of sci-fi movies since then, with 1977's *Empire of the Ants*, 1995's *Mosquito*, and 2002's *Eight-legged Freaks*.

# Not of this World: Insects as Aliens

Many sci-fi films over the last couple of decades have featured aliens based on arthropods, or utilize alien behavior based on that of arthropods. Arthropods are inherently strange and "alien-looking" creatures for most people. Aliens based on them, rather than, say humans, will therefore be more foreign and harder for viewers to relate to, enhancing the exotic mood of a film. In *Alien* (1979), the other-worldly antagonist mimics the lifecycle of a parasitoid in that it incubates in the chest of Kane before bursting free, killing him in the process. In 1997's *Men in Black*, the main antagonist is an alien "bug" loosely based on the Madagascan hissing cockroach that inhabits a human's skin before ultimately bursting free at the climax of the movie as the giant he was meant to be. Since this particular alien is based on the almost universally reviled cockroach, when he explodes from within, it's humorous rather than terrifying. *Starship Troopers* (1997) also features aliens based on arthropods. The troopers confront armies of ant-like "arachnoids" and the occasional giant, flame-throwing beetle (loosely based on the real-life bombardier beetle). In 2013's *Ender's Game*, the alien race Earth is warring with is loosely based on ants with queen-led colonies and a generally arthropoid appearance. They're even called "formics," which is based on Formicidae, the family to which ants belong.

# I Don't Feel So Good: Bad Insect-Human Interactions

Insects are alien, so what could be worse than turning into an insect? Several sci-fi movies confront this very question, and the clear classic in this genre is 1958's *The Fly*. This is a classic mad-scientist film where the protagonist, working on teleportation, accidently blends himself with a fly trapped in the teleportation device. He emerges with the head and "arm" of a fly, while the fly disappears with his head and arm. The film was remade in 1986 with Jeff Goldblum and a perhaps better understanding of genetics. Both movies had sequels that continued the trope. *The Wasp Woman* (1959) told the story of a woman who misuses an anti-aging product refined from the queen jelly of wasps and becomes a violent hybrid. Many other science-fiction movies have exploited this plot device.

# When Good Go Bad: Real Insects as Movie Villains

Most of the movies we've discussed so far rely on the special effects shop to produce the insects depicted in them, but a few movies have attempted to use real, live insects. One of the first of these was *The Naked Jungle* (1954), in which a South American plantation is threatened by a huge swarm of army ants. The movie had numerous scenes of teeming hordes of real ants (mostly leafcutters, I think), and in one segment, had them making rafts of leaves to cross a canal (ants do many remarkable things, but they don't, as far as I know, make boats). Another movie that used real insects was 1985's *Creepshow*, an anthology movie based on a television show with one segment telling the story of the invasion of a rich, malicious, germaphobic recluse's apartment by hordes of cockroaches. Real roaches were used to more comic effect in *Joe's Apartment* (1996). The live roaches were supplemented by animated, talking roaches in this tale of a likeable slob trying to get the girl and defeat the criminals.

A special category of the "real insects as villains" movie is the "killer bee" movie. The first of this genre was *The Deadly Bees* (1966), a British film actually inspired by a novel written two decades earlier. In this movie, two beekeepers on an isolated island compete for the attentions of a young woman. One has managed to create a race of "killer" bees he can manipulate with pheromones and sound (there are some real problems with some of the biology in this movie). The accidental introduction of Africanized bees into the Americas in the mid-1950s inspired a number of (mostly bad) movies about malevolent hordes of attack bees. Indeed, there have been at least three movies titled simply *Killer Bees* (1974, 2002, 2008)! There are at least eight or nine other killer bee movies in the catalog of horror movies.

There are several other subgenres of arthropod horror movies—the spider movie (normal-sized or huge), the scorpion movie, the evil giant crustacean movie. As long as most of the population finds creatures with exoskeletons creepy, alien, and somewhat intimidating, insects and other arthropods will continue to appear in movies as creatures intended to haunt your dreams.

# Real Insects Being Real Insects

The last genre we'll discuss is perhaps the smallest, but in some ways, the most interesting collection of theatrical movies featuring insects. In this genre, real insects are depicted being real insects—no mass attacks, no special effects. There are only a few movies of this type, and the two most prominent provide an interesting contrast in how insects may be exploited in a theatrical production. The earlier of these two is *The Hellstrom Chronicles* (1971), a quite dark movie featuring a fictional scientist who lays out the case that the insects will ultimately dominate the earth over us in spite of all our technology. The movie has spectacular natural history footage of insects, but these images are accompanied by ominous music and the solemn narration of Professor Hellstrom. The other movie is quite different in tone. *Microcosmos* (1996) is a beautifully filmed movie that describes the goings-on in a French meadow during the summer. There is virtually no narration, and the images are accompanied by some lovely music and the sounds of the creatures themselves. Different segments of the movie very effectively convey humor, romance, combat, and even the frustration one experiences when stuck in a traffic jam. The movie received numerous awards and did quite well in its rather limited theatrical run. One can hope that more movies of this sort will come out in the future!

# GLOSSARY

**Accessory gland-** Paired glands found in the reproductive tracts of both male and female insects, although their functions differ in the two sexes. In males, they produce substances to support the sperm, and perhaps nuptial gifts; in females, they make the glues and other substances the female uses to secure her eggs. In stinging insects, the accessory glands make venom.

**Aedeagus-** The copulatory organ of an insect, analogous to the penis in mammals.

**Africanized honey bees-** Honey bees demonstrating the aggressive nature and other undesirable characteristics of African honey bees. Dominantly inherited when these bees mate with European strain honey bees. First released (accidentally) in Brazil in 1956; now found in southern border region of the United States.

**Agent-** A living organism causing disease in arthropod –vectored disease cycles.

**Air sacs-** Unreinforced areas in trachea that allow the insect to force air in and out of the respiratory system.

**Allomone-** Chemical communication between organisms, which is positive for the sender and negative for the receiver.

**Ametabolous-** Synonymous with no metamorphosis. Found in most primitive, wingless insects. Young basically look like miniature adults; adults distinguished by having functional sex organs.

**Ammonia-** A toxic waste product produced when animals metabolize protein. Aquatic insects can get rid of it easily by producing large amounts of dilute, ammonia-laden urine. Terrestrial insects convert it to uric acid to save water.

**Anaphyllaxis-** A severe allergic response in a vertebrate. In the worst case of "anaphylactic shock," death may result.

**Annual migration-** Migration that occurs every year at a predictable time.

**Anoplura-** The order of insects containing the sucking lice.

**Antenna (pl. antennae)-** The primary chemical reception ("smell") sensors of an insect. Insects have one pair; the main part of the antenna, and the part furthest from the body, is the "flagellum." All the muscles that move the antenna are located at its base.

**Apostrophe-** In poetry, a poem made as an address to an absent person or thing, including an insect. "To a …"

**Apterygota-** Primitively wingless insects. No metamorphosis.

**Arachnida-** The class containing the spiders, ticks, and mites, among many other diverse forms. Arachnids have two major body regions, eight legs, and no antennae. Most are predators or parasites of other animals.

**Arthropods-** Animals, in the phylum Arthropoda, with bilateral symmetry, chitonous exoskeletons, segmented bodies, jointed appendages.

**Asexual reproduction-** Reproduction that doesn't necessarily require mating.

**Asynchronous muscle-** An insect muscle that contracts multiple times in response to a single nerve impulse.

**Augmentation biological control-** Biocontrol reliant on releasing large numbers of biological control agents to suppress a pest population. Using biocontrol agents as "living insecticides."

***Bacillus thuriengensis-*** Soil microbe found virtually worldwide that produces crystalline proteins toxic to various insects. Widely used as a microbial insecticide, and source of genes for transgenic crop plants.

**Basement membrane-** Non-living layer that protects the epidermis from the hemolymph.

**Binomial system of nomenclature-** Our two-name system for naming species of living things – the Genus and the species within it. The monarch butterfly's binomial scientific name is *Danaus plexippus*. Always italicized, with first name capitalized and second name not.

**Biological Control-** Using living organisms, the "natural enemies" of pests, to reduce pest populations.

**Biological disease transmission-** The case in which the agent causing disease must pass some part of its lifecycle within the vectoring organism.

**Bioluminescence-** Light produced by a living organism.

**Bisexual reproduction-** Reproduction requiring mating between a male and female parent.

**Bubo-** Characteristic, gross swelling of lymph nodes associated with plague infection.

**Carnivorous plants-** Plants that harvest insects and other arthropods to meet their demands for nitrogen and other nutrients.

**Castes-** Different body plans for different jobs, within the same species of eusocial insect.

**Cattle grub-** Dipteran pests that develop under the skin of cattle, damaging leather and causing scar tissue in meat.

**Cerci-** Paired appendages near the end of insect's abdomens that usually have sensory function (much like antennae on the butt!) but are often modified for other functions.

**Chilopoda-** The class containing the centipedes, arthropods with many-segmented bodies, one pair of legs per segment, one pair of antennae, flattened bodies, and venomous fangs. All are predators.

**Chitin-** The poly-saccharide ("many sugared") compound that forms the structural framework for the arthropod exoskeleton. Composed of repeating, linked molecules of glucosamine.

**Chloroquine-** Synthetic antimalarial drug.

**Chorion-** The shell of an insect egg. Made of protein.

**Class-** The divisions of life within phyla. Insecta is a class within Arthropoda.

**Cochineal-** Brilliant red dye produced by scale insects found on prickly pear cacti.

**Commercial agriculture-** Agriculture in which farmers produce crops primarily for sale to others.

**Communication-** Exchange of information between two living organisms.

**Communal insects-** Species that aggregate during some phase of the lifecycle but do not maintain social contact throughout the lifecycle.

**Compound eye-** A visual sensory organ composed of many smaller subunits called ommatidia or "facets." Forms a mosaic image with resolution dependent on the number of ommatidia.

**Conservation biological control-** Controlling pest insects by protecting the biological control agents already present.

**Crustacean-** An arthropod with a calcium-reinforced chitinous exoskeleton, two major body regions, two pairs of antennae, and, usually, five pairs of legs. Most are marine (living in the ocean) or aquatic; the class includes many other diverse forms. Now thought to be ancestral to insects.

**Cultural control-** Pest management that relies on modifying something we would do any way to help reduce pest populations. In agriculture, could include modifying planting or harvest dates, tillage changes, crop rotation, etc. In other circumstances, sanitation is an important cultural control strategy.

**Cultural Entomology-** The study of the impact of insects on human culture and society.

**Cursorial-** Adapted for running.

**Cuticle-** The non-living parts of the insect exoskeleton. Composed of three layers: endocuticle, exocuticle, and epicuticle.

**Cuticle poison-** An insecticide that compromises the integrity of the insect exoskeleton, through either abrasion (dusts), interfering with epicutiucular waxes (soaps), or interfering with the insect's ability to synthesize chitin. Generally very safe for non-target organisms.

**Delusional parasitosis-** A condition in which the sufferer believes non-existent insects inhabit one's person.

**Dengue-** Viral disease transmitted by *Aedes* mosquitoes. Also known as Breakbone fever.

**Diapause-** A resting state that insects enter to avoid inhospitable environmental conditions.

**Diploid-** The normal number of chromosomes in most cells. Diploid cells have a pair of each chromosome; one member of each pair came from the animal's mother, the other from its father.

**Diplopoda-** The class containing the millipedes, arthropods with many-segmented, cylindrical bodies, two pairs of legs per segment, one pair of antennae, and calcium in their exoskeletons. Most are plant feeders or scavengers, and many have potent chemical defenses.

**Direct flight muscles-** Major flight muscles directly attached to the bases of the wings.

**Disease agent-** An organism causing negative effects in another organism.

**Dorsal vessel-** The pump in the insect's circulatory system. Consists of the "heart," a perforated, muscular tube in the abdomen, and the "aorta" —basically a hose that conducts hemolymph forward to the head.

**Economic Injury Level-** The pest population level at which the value of lost crop yield **EQUALS** the cost of controlling that pest population.

**Economic Threshold-** The population level of a pest where action must be taken to keep the population from exceeding the Economic Injury Level.

**Ectognatha-** Hexapods with exposed mouthparts.

**Ejaculatory duct-** The tube that conducts sperm from the male reproductive tract into the female: passes through the aedeagus.

**Elytra-** The hard, shell-like for wings of beetles.

**Endocuticle-** The innermost layer of the cuticle. Composed of chitin and protein; flexible but not stretchy. Found throughout exoskeleton.

**Endopterygota-** Insects whose wings develop out of sight: those insects that undergo complete metamorphosis.

**Entognatha-** Primitive insect-like animals with mouthparts hidden in the head.

**Entomophagy-** The practice of eating insects.

**Entomophobia-** An irrational fear of insects.

**Epicuticle-** The outermost and thinnest layer of the cuticle. Composed of waxes, cements, protein; responsible for waterproofing the exoskeleton. Found on all external portions of exoskeleton.

**Epidemic typhus-** Bacterial disease caused by *Rickettsia prowazekii* and transmitted by the body louse. Outbreaks often associated with times of conflict and war; historically, the course of conflicts was often regulated in part by the disease.

**Eusocial insects-** Species with cooperative brood care, reproductive division of labor, and perennial colonies with overlapping generations.

**Exocuticle-** The middle layer of the cuticle; composed of chitin, protein, and phenolic compounds which give it hardness and rigidity. Found in the exoskeleton wherever strength, hardness, and rigidity are required.

**Exopterygota-** Insects that develop wings on the outside: those that undergo incomplete metamorphosis.

**Exoskeleton-** The external support structure of arthropods. In insects, mostly chitin and other stuff. Composed of the living, single-cell thick epidermis and the three-layered cuticle. Also lines the foregut, hindgut, and major trachea of the respiratory system.

**Exotic pest-** A pestiferous organism not native to the area.

**Family-** The divisions within orders of living things. The monarch butterfly is in the family Danaidae within the order Lepidoptera.

**Fat body-** A storage organ in the insect's body. Well-nourished insects are full of fat body. Also converts nitrogenous waste to uric acid and plays a roll in hormonal control.

**Fore-gut-** The front part of the insect digestive system. Begins with the mouth and ends at the cardiac valve. Lined with cuticle. Contains crop, which stores excess food for later digestion.

**Forensic entomology-** Using insects to understand crime scenes.

**Fossorial-** Adapted to digging in soil.

**Ganglion-** Cluster of nerve cell bodies.

**Genus (plural genera)-** The divisions of life within families. The monarch butterfly is in the genus *Danaus* within the family Danaidae. The Genus name is always capitalized and always italicized.

**Gills-** In insects, outgrowths of the tracheal respiratory system that increase the area for gas exchange in some aquatic insects.

**Halteres-** Highly modified wings used for balance rather than flight. Usually are club-shaped structures; found on the metathorax of true flies (Diptera).

**Haplo-diploidy-** The sex determination system in Hymenoptera and some other animals, in which a fertilized egg produces a female offspring, while an unfertilized egg produces a male offspring.

**Haploid-** Half the normal number of chromosomes for a given species. Generally only eggs and sperm are haploid; they only contain one of each chromosome.

**Heartworm-** Deadly disease of domestic dogs and other animals caused by a roundworm and transmitted by mosquitoes. Found throughout North America.

**Hemelytra-** The half-leathery, half-membranous wings of true bugs like stink bugs.

**Hemimetabolous-** synonymous with incomplete or partial metamorphosis. Nymphs look like miniature adults, but with developing wing pads rather than functional wings. Nymphs and adults typically have same kinds of mouthparts, feed on the same things, and live in the same places, so they compete with each other.

**Hemocytes-** Blood cells, primarily responsible for defense against disease and parasites.

**Hemolymph-** The blood of insects. Contains water, defensive cells, nutrients, hormones, etc., but does not transport oxygen in most insects.

**Hemocoel-** The body cavity of the insect, generally full of hemolymph bathing all the internal organs and muscles.

**Hexapod-** An arthropod with six legs.

**Hindgut-** Back end of insect digestive tract. Lined with cuticle; packages waste for elimination and reabsorbs water and salts from waste before disposal. Starts at the pyloric valve and ends at the anus.

**Holometabolous-** Synonymous with complete metamorphosis. Four life stages: egg, larva, pupa, adult. Larvae and adults often bear no resemblance whatsoever to each other. Larvae and adults may have radically different mouthparts and live and feed in very different places.

**Homeothermic-** Able to maintain a constant body temperature under variable environmental temperatures. See Poikilothermic.

**Hormone mimic-** An insecticide that imitates the action of an insect hormone. Juvenile hormone mimics keep insects immature until death; molting hormone mimics cause premature molting and death.

**Horse bot-** Any one of three species of Dipteran pests of horses that complete their larval development in the stomach of horses.

**Host plant resistance-** Breeding crop plant varieties that reduce the impact of pests on yield and quality.

**Hypopharynx-** A structure in the insect's mouth that functions much like a tongue.

**Importation Biological Control ("Classical Biocontrol")-** Importing exotic natural enemies for exotic pest species. Vedalia beetle imported to California to save the citrus industry from the cottony cushion scale is the classic example of a successful program.

**Indirect flight muscles-** Major flight muscles attached to the walls of the thorax rather than the bases of the wings.

**Insecta-** Arthropods with three major body regions, six legs, one pair of antennae, and, in most species, winged adults. The only flying invertebrates and the dominant life form on the planet in terms of diversity.

**Insecticide-** Pesticide developed for use against arthropod pests.

**Insect pathogens-** Disease organisms that kill or sicken insects but not other organisms. Valuable biological control organisms.

**Instar-** A stage in the life cycle of an insect. When an insect molts, it goes from one instar to the next; the adult is the final "perfect" instar, since it has functional sex organs and wings.

**Juvenile hormone (JH)-** A chemical produced indirectly by the insect's brain that tells the epidermis what kind of cuticle to make when it next molts. High JH and high MH result in another immature cuticle; MH without JH results in an adult cuticle. In insects that go through complete metamorphosis, high MH with a little JH results in a pupal cuticle.

**Kairomone-** Chemical communication between organisms that is negative for the sender and positive for the receiver.

**Keratin-** Fibrous structural protein found in the skin, fur, feathers, horn, scales, and other structures covering vertebrate animals. Subject to attack by clothes moth larvae and skin beetles.

**Kingdom-** The largest division of life on earth. Insects are in the Kingdom Animalia.

**Labium-** The "bottom lip" of an insect. Often have sensory, leg-like palps which work much like the palps on the maxillae.

**Labrum-** The "upper lip" of the insect's mouth.

**Larva-** The immature stage of insects that go through complete metamorphosis. Never have anything that looks like wings, no compound eyes, and often very different from adults.

**Linnaeus-** The father of the modern classification of life. He formalized the hierarchal system of classification ("King Phillip Cried Oh For Goodness Sakes!"—Kingdom, Phylum, Class, Order, Family, Genus, Species) and the binomial system of nomenclature.

**Locust-** A large, often migratory, grasshopper.

**Luciferin-** The chemical that when combined with ATP and the enzyme luciferinase, produces light in insects.

**Malaria-** Disease vectored by *Anopheles* mosquitoes, caused by protozoan parasites in the genus *Plasmodium*. Single greatest killer of humanity.

**Malphigian tubules-** The "kidneys" of the insect, responsible for collecting and concentrating metabolic waste in the insect. Join the digestive tract at the beginning of the hindgut.

**Mating disruption-** Using pheromones to confuse pest insects so that they can't find each other to mate.

**Mandibles-** The "jaws" of the insect. Found behind the labrum and usually the hardest structures in the body of an insect with chewing mouthparts.

**Maxilla (pl. maxillae)-** Leg-like appendages in the mouth of insects that help them manipulate their food and often have taste receptors. Generally behind the mandibles.

**Mechanical control-** Insect pest control that uses a purpose-built mechanical device or hands (and feet).

**Mechanical disease transmission-** The case where a disease agent is transmitted on the surface or in the mouthparts of the vectoring organism.

**Median filament-** An extension of the top plate on one of the last abdominal segments that looks like a long tail. Generally found only on some primitive insects and probably sensory.

**Meiosis-** The reduction division of cells in the reproductive organs that produces haploid sperm or eggs.

**Membranous-** Filmy and often transparent. Membranous insect wings often look like they are made of cellophane wrapping.

**Micropyle-** The tiny hole in the chorion of an insect egg allowing sperm to enter and fertilize the ova (egg).

**Midgut-** Middle part of digestive system. Where digestion and absorption of nutrients occurs. Only part of gut not lined with cuticle. Ends at Pyloric valve.

**Migration-** Directional movement of large numbers of organisms.

**Molting fluid-** The substance secreted by the epidermis that recycles the old endocuticle when the insect is preparing to molt.

**Molting hormone (MH)-** A chemical produced indirectly by the insect brain that signals the insect epidermis to begin making a new exoskeleton and recycle parts of the old.

**Monoculture-** Agricultural production (or storage) of a single crop (or livestock animal) over a large area. May also be represented by a concentration of an agricultural resource.

**Myiasis-** Parasitic infestation of living vertebrate tissue by larval flies.

**Natatorial-** Adapted to swimming.

**Nerve poison-** Insecticide with primary activity on the insect nervous system. Most modern insecticides are nerve poisons; most newer ones are quite safe for non-target organisms, but some older ones are quite toxic.

**Nomadic migration-** Migration in any direction in response to the availability of resources.

**Nymph-** The immature stage of an insect that goes through incomplete metamorphosis. Often quite similar to adults, with similar mouthparts; often have compound eyes and visible wing pads that get larger with each successive molt.

**Nuptial gift-** Product or object provided to female (usually) by a male (usually) to persuade her (usually) to mate.

**Ocellus (plural ocelli)-** "Simple eyes" that don't form complex images but can detect color, light, and dark. Most insects probably use them to help keep track of time and season.

**Onychphoran-** A phylum of creatures that may be ancestral to insects. They have many unjointed legs. *Peripatatus* is a living example of an Onychphoran.

**Ootheca-** A protective structure many female insects construct around their eggs. A product of the female's accessory glands. A mantis's has the texture of Styrofoam; a roach's looks like a tiny purse.

**Order-** Divisions of life within classes. Lepidoptera is the order within the class Insecta containing the moths and butterflies.

**Ovary-** The primary sex organ of females, responsible for making haploid eggs. In insects, the ovaries are composed of a number of ovarioles, each of which functioning like a small egg assembly line.

**Oviducts-** The tubes that conduct eggs away from the ovaries.

**Oviparous-** Lays eggs that hatch some time later.

**Ovipositor-** An appendage found on the abdomens of female insects that they use to lay eggs. In some Hymenoptera (bees, ants, and wasps), the ovipositor is modified into a defensive stinger.

**Ovoviviparous-** Produces eggs but holds them internally until they are ready to hatch.

**Paedeogenesis-** Parthenogenic reproduction by an immature form; happens in some insect larvae.

**Parasite-** An organism that lives in, on, or near another organism at that organism's expense.

**Parasitoid-** A parasite that kills its host as part of its life cycle. Most important insect parasitoids of pest insects are wasps (Hymenoptera) and flies (Diptera).

**Parthenogenesis-** Reproduction without mating. Usually results in clones of the mother.

**Pathogen-** An organism causing disease in another.

**Pesticide-** An agent or chemical used to kill or reduce the impact of some organism we regard as pestiferous.

**Pheromone-** chemical that facilitates communication between members of the same species.

**Phylum-** The great divisions of life within kingdoms. Insects are in the Phylum Arthropoda.

**Physical control-** Insect control strategies that rely on modifying the pest's environment to make it less suitable for the pest. Could include temperature manipulation or atmosphere manipulation.

**Plague-** Bacterial disease caused by *Yersinia pestis*, vectored by fleas; the reservoir is generally rodents. Also known as the "Black Death," had grave consequences for medieval European civilization.

**Plastron-** In insects, an air bubble trapped by hydrophobic (water repelling) setae in some aquatic insects which enhances gas exchange.

**Poikilothermic-** Having a body temperature roughly that of the surrounding environment. See homeothermic.

**Pollination-** Transfer of pollen (male germ cells) from one flower to the style and ovary of another flower, accomplishing fertilization.

**Polyembryony-** The condition where a single egg splits into many embryos, producing that many clones of the original egg.

**Predator-** A free-living organism that captures and consumes animals for its nourishment.

**Proboscis-** A feeding structure evolved from the chewing mouthparts of ancestral insects found in some modern insects that feed on liquid diets like nectar.

**Pronotum-** The top plate on the prothoracic segment. Often highly modified for defense, mating contests, or camouflage purposes.

**Pterygota-** Winged insects.

**Pupa-** The stage in complete metamorphosis between the larva and the adult. Generally not mobile, doesn't eat, doesn't excrete. In this stage everything that was the larvae transforms in to the adult.

**Quinine-** Anti-malaria drug made from bark of the tropical cinchona tree. Found in tonic water.

**Raptorial-** Adapted for catching prey animals.

**Reservoir (disease)-** An organism which sustains a disease agent on the landscape for long periods of time. Reservoirs are usually unaffected, or minimally affected, by the agent.

**Resilin-** A rubbery protein that stores energy and provides some stretch in parts of the insect exoskeleton. Also stores energy in some insect muscles.

**Respiratory poisons-** Insecticides that compromise the insect's ability to breathe. Oil suffocants are generally pretty safe; fumigant gases can be quite hazardous.

**Salivary glands-** glands located near mouth that provide lubrication for swallowed food; sometimes modified to produce venom, anticoagulants, and, in caterpillars, silk.

**Saltatorial-** Adapted to jumping.

**Screwworm (Primary screwworm)-** Dipteran pest of livestock responsible for millions of cattle and other livestock deaths. Eradicated from North and Central America through sterile insect release program.

**Semisocial insects-** Species with cooperative brood care and reproductive division of labor.

**Seta (pl. setae)-** The "hairs" found on insects. Chitinous outgrowths of the exoskeleton.

**Shellac-** Durable, high gloss furniture finish produced by an Asian scale insect.

**Skeletal muscles-** Those that move body parts of an insect.

**Silk-** Commercial product of the domesticated silkworm, *Bombus mori*. Produced by the salivary glands of the caterpillar as it spins its cocoon.

**Simile-** In poetry, a comparison of unlike objects.

**Solitary insects-** Species that don't interact with others of their species except to mate or in response to resources.

**Species-** A population of organisms capable of producing viable offspring and reproductively isolated from similar populations. The scientific name of a species always consists of two Latinized words. The first is genus to which it belongs and the second is the name of the species within the genus. The genus is always capitalized; the species name is not.

**Sperm-** Male germ cells. Haploid, and produced by the testes.

**Spermatheca-** A structure in the reproductive tract of female insects that stores sperm until they are needed to fertilize eggs.

**Spiracles-** The holes along the sides of the abdomen and some thoracic segments that open to the tracheal respiratory system.

**Sterile Insect Release Technique-** An insect eradication technique in which sterile male insects are released (in huge numbers), resulting in matings with females which produce infertile eggs. Continued releases eventually drive the pest population to extinction.

**Stridulation-** Sound produced by dragging a row of peg-like structures across a file-like structure.

**Stomach poison-** An insecticide that the insect must eat in order to be intoxicated. Most of these older insecticides are no longer used; many contained arsenic and other dangerous elements.

**Styli-** Peg-like appendages found along the sides of the abdomen in some very primitive insects like silverfish.

**Subsistence agriculture-** Agriculture in which the farmers grow enough food for them and their families but little more.

**Subsocial insects-** Species in which a single female or a mated pair care for discreet broods of immatures.

**Surgical maggots-** Fly larvae, raised in sterile conditions, used to remove necrotic (dead) tissue from wounds.

**Synchronous muscle-** Muscle that contracts one time for a single nerve impulse

**Target-** In arthropod vectored disease cycles, the host organism sickened by the disease agent.

**Testes-** Primary male sex organs, make sperm.

**Thorax-** The middle portion of an insect body and the center of locomotion. Always has three segments with, usually, one pair of legs per segment. The segment closest to the head is the prothorax; the middle, the mesothorax, and the hindmost, the metathorax. Wings may be found on the meso and meta but never on the prothroacic segment.

**Thysanura-** The primitively wingless insect order containing silverfish and their relatives. Three tail appendages, scaled bodies, can molt as adults (unlike virtually all other insects).

**Trachea-** The large respiratory pipes of the insect respiratory system, opening to the outside through the spiracles. Lined with cuticle.

**Tracheole-** The tiny pipes that conduct gases to every living cell in the insect's body from the larger tracheal elements.

**Transgenic crop plant-** A crop variety modified through genetic engineering technology to contain genes from other species yield desirable traits. Many crop varieties have had genes from *Bacillus thuriengensis* inserted to yield resistance to caterpillar and/or beetle pests.

**Trilobite-** Ancient, extinct arthropods which dominated the ocean between 600 and 400 million years ago; distant ancestors of insects

**Trophyllaxis-** Sharing of gut contents between members of the same species, either through regurgitation or defecation.

**Tymbal (or tympanum)-** In insects, a thickened plate of chitin connected to powerful asynchronous muscles; deflection of the tymbal by the muscle produces sound. Typified by cicadas.

**Uric acid-** The nitrogenous compound insects form to get rid of nitrogen produced as waste product of protein metabolism. Relatively nontoxic and solid, it conserves water in the process of getting rid of this metabolic waste.

**Urticating seta ("hairs")-** Stinging setae found on some insects, including some caterpillars

**Vas deferens-** The tube that conducts sperm away from the testes.

**Vector-** An organism transmitting a disease agent from an infected to an uninfected host.

**Vesicle ("seminal vesicle")-** An organ that stores sperm in the male reproductive tract.

**Vestigial-** Structures which no longer function for their original purpose. Many adult insects have vestigial mouthparts since they don't feed as adults.

**Visceral muscles-** Those that move the internal organs of an insect.

**Viviparous-** Gives birth to live young nourished by the mother before birth.

**Yellow fever-** Viral disease transmitted by *Aedes* mosquitoes. Nearly stopped construction of Panama Canal; halted Leclerc's invasion of Haiti.